THE LONG
HAUL PIONEERS

THE LONG HAUL PIONEERS

A celebration of Astran, leaders in overland transport
to the Middle East for over 45 years

Ashley Coghill

Epitaph for a Truckie

A truckie stood at the pearly gates, his face was worn and old.
He meekly asked the man of fate for admission to the fold.
'What have you done?' St. Peter asked, 'to seek admission here?'
'I was driver of a Scania road-train for many, many a year.'
The gates swung open sharply as St. Peter touched the bell.
'Come in my boy and take a harp. You've had enough of hell.'

First published 2010, reprinted 2010

Copyright © Ashley Coghill, 2010

The moral right of the author in this work has been asserted

All rights reserved. No parts of this publication may be reproduced, stored in a retrieval system, or transmitted, in any form or by any means electronic, mechanical, photocopying, recording or otherwise, without prior permission of Old Pond Publishing.

ISBN 978-1-906853-33-4

A catalogue record for this book is available from the British Library

Published by
Old Pond Publishing Ltd
Dencora Business Centre, 36 White House Road
Ipswich IP1 5LT United Kingdom

www.oldpond.com

Book design and layout by Liz Whatling
Cover design by Chris Stringer and Liz Whatling
Printed and bound in China for Latitude Press

This book is dedicated to Caroline Paul
In loving memory

The Long Haul Pioneers

Main truck routes through Europe

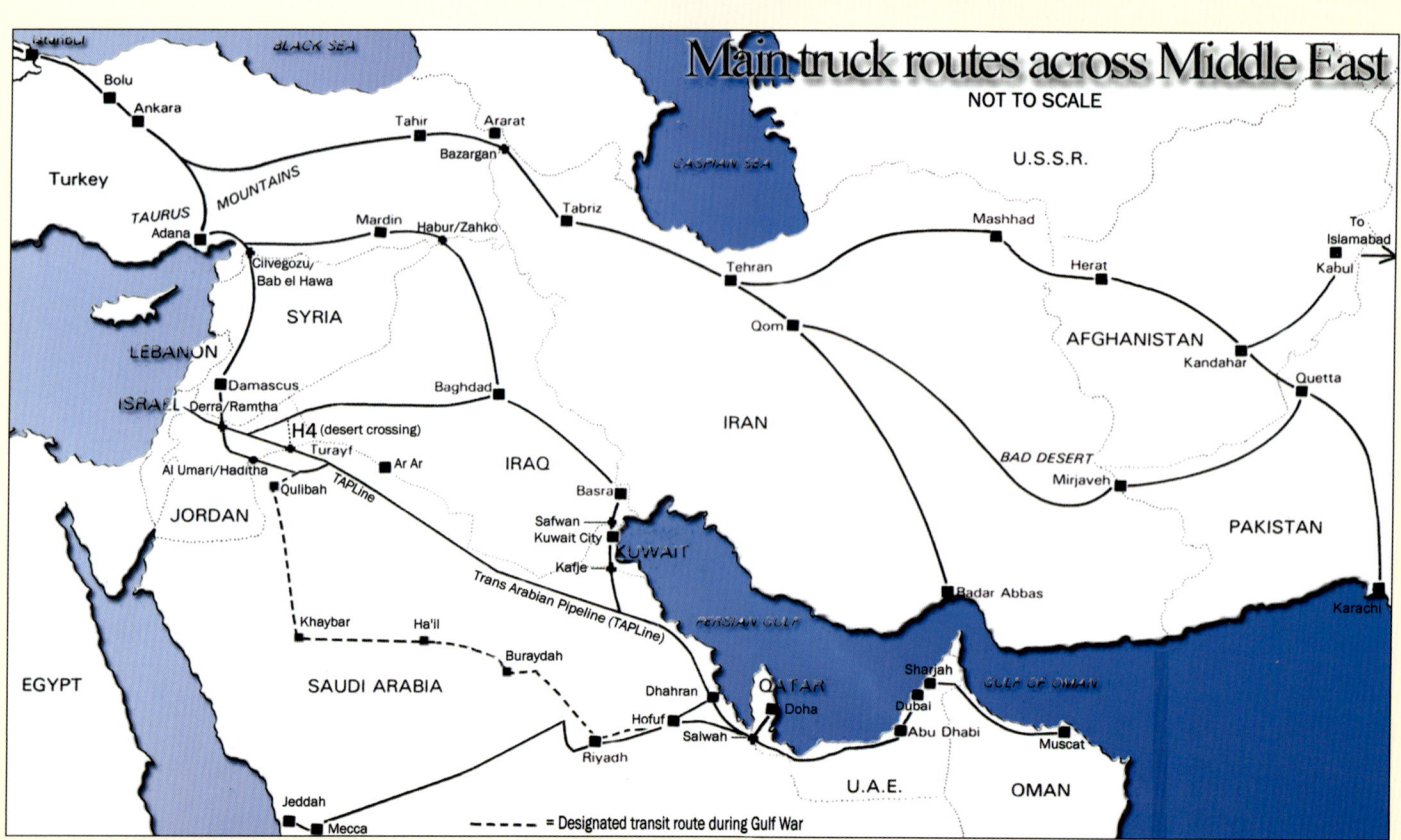

Main truck routes across Middle East

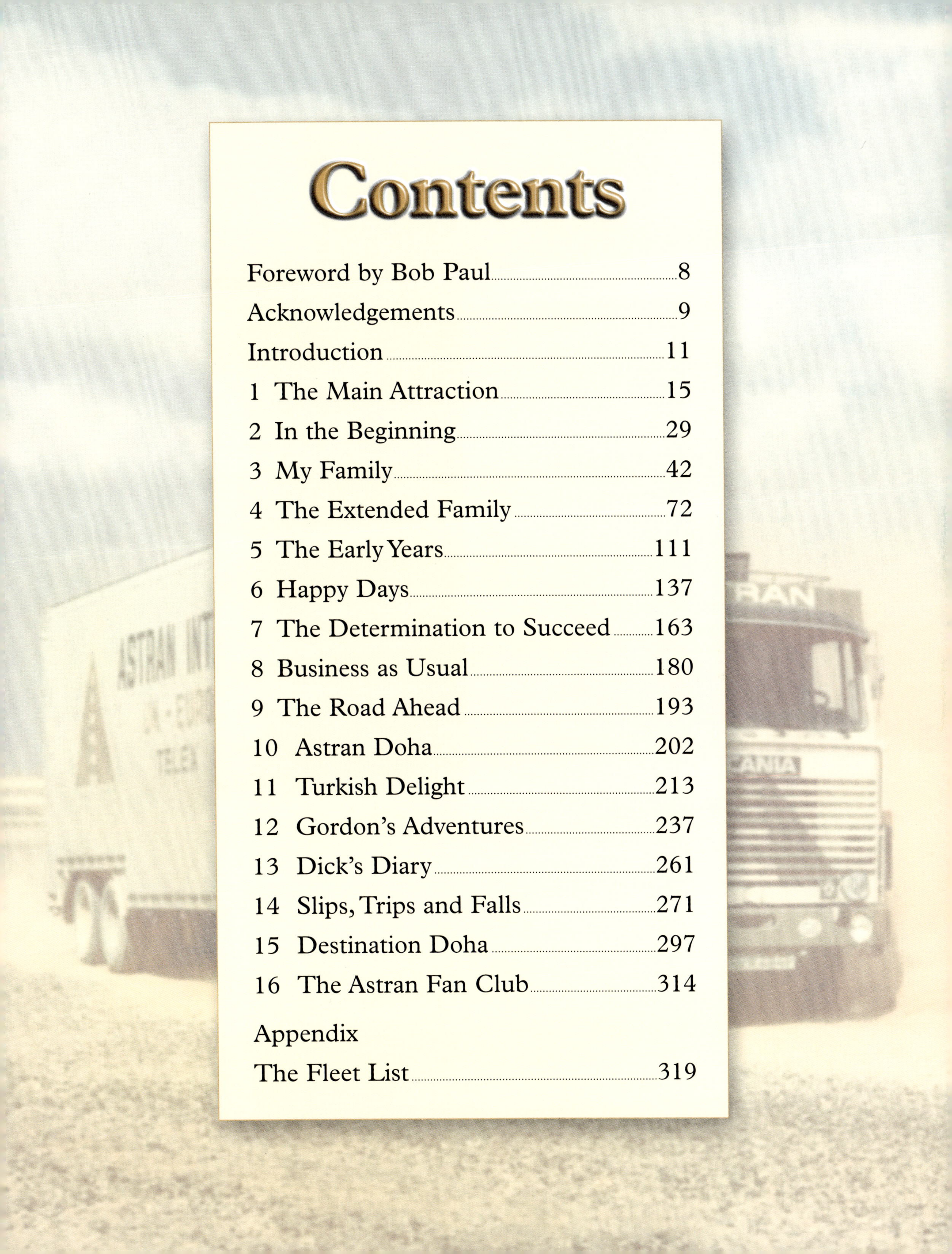

Contents

Foreword by Bob Paul 8

Acknowledgements 9

Introduction 11

1 The Main Attraction 15

2 In the Beginning 29

3 My Family 42

4 The Extended Family 72

5 The Early Years 111

6 Happy Days 137

7 The Determination to Succeed 163

8 Business as Usual 180

9 The Road Ahead 193

10 Astran Doha 202

11 Turkish Delight 213

12 Gordon's Adventures 237

13 Dick's Diary 261

14 Slips, Trips and Falls 271

15 Destination Doha 297

16 The Astran Fan Club 314

Appendix

The Fleet List 319

Foreword

With new buildings and roads being constructed almost every day, it is very easy to forget how things used to be. It is the same with road transport as modern technology makes everything effortless and takes away all the traditional methods and values that it once had. While I agree that we should never stand in the way of progress, I also believe we should never forget the past.

I have spent over forty years in the transport industry and have experienced many changes during my time as a driver and company director. It has been my life, but I have never regretted any of it and I never for one minute missed the medical career that I yearned for as a boy and then trained for as a young man.

Who could ever have imagined that when Mike Woodman and I took our first precarious steps as lorry drivers in 1964, we would be the ones responsible for opening the gates and allowing thousands of trucks to follow us across Europe and into the Middle East?

For those of us who were privileged to drive through strange and mystical countries, we still have precious memories and clear images in our minds, but today's younger generation will have no idea of how things were, so they will surely be curious and fascinated to learn about an era of road transport which will never be repeated.

The best times I ever had were during those early years when we were fighting for something, exploring new places and finding it all very exciting and adventurous. When we got to the other end and only after we had struggled against all the odds, we knew we had achieved something special. It was sheer determination that made it all worthwhile. Working and living so closely with the men I called 'my boys' made it a unique camaraderie. If we hadn't been so close, I think a lot of the men would have fallen by the wayside. They were so dedicated and without them we wouldn't have got anywhere.

Now, thanks to the hard work of Ashley Coghill and his relentless quest to locate my old employees, he has been able to document and illustrate a unique lifestyle which wasn't just a job; it was a way of life.

I have enjoyed being involved with Ashley's research and it has rekindled my interest in something that I was very passionate about. The company which I built up and was part of for so many years is still trading today, albeit under new ownership, and it is wonderful to see the complete history put together like this.

I hope that anyone reading this book will become immersed in the conditions and traumas that we all had to endure in the early days and I hope it offers an understanding and an insight into something that most people never dreamed of doing.

Bob Paul with author Ashley Coghill.

Bob Paul, 2009

Acknowledgements

To produce a factual book with an in-depth historical and photographic theme is no easy task. It requires an awful lot of research, but perhaps more importantly it requires total dedication and a great deal of passion for the subject. Luckily, I have oodles of both.

In writing this book I have been totally dependent on the people who worked for Asian Transport and Astran over the years. I have searched relentlessly not only for the employees, but for numerous other people with connections to the company and it would be impossible to name every one of them who has provided photographs, information and advice. I hope they will feel that this book adequately reflects their help.

I have also received a special amount of help and support from certain individuals without whom this book would never have come to fruition but I have found it extremely difficult deciding who the most significant person has been. Nevertheless, there are two people who deserve a very special mention before anyone else.

Firstly my partner, Dawn. Thank you for allowing me the freedom and time to pursue this project from start to finish. You have never complained when I have gone off on yet another jaunt to interview one of the men and you have never moaned when I have sat for hours upon end typing up my research. At times you have been an author's widow, but without realising it. By allowing me to do all of that, you have helped me more than you could imagine. Thank you so much for your patience and understanding.

Secondly, my sincere thanks must go to Mr Bob Paul. You have no idea how much your help has meant to me. Your knowledge of Asian Transport and Astran is quite simply remarkable and listening to your crystal-clear recollections and experiences has been amazing. I am so grateful to you for spending so much of your time with me and for allowing me to delve into your past. I feel extremely honoured and privileged to have met you and to have had the opportunity to get to know you.

When I decided to start this project, I already had some background information and had been introduced to Asian Transport's very first driver, Gordon Pearce, so I was able to start the ball rolling with his help. His encouragement from the outset has been second to none; whenever I have needed help and advice he has always been there with first-hand knowledge which has proved invaluable. He has also spent many hours jotting down his adventures for me to elaborate on and, apart from his tremendous help, I know he has thoroughly enjoyed being involved. Gordon, I cannot thank you enough.

Everyone to a man and woman whom I have contacted during my research has given me their time freely, detailing their experiences and showing me their precious photographs and documents. Not one of them has ever turned me down for a phone call or a visit so I would like to express my gratitude and appreciation to you all. I would especially like to thank Peter Cannon for putting up with my relentless phone calls and for taking the time to read through my work, when I am sure he would rather have been doing something far more important.

Without everyone's help, this book would be nothing. It is filled with their memories, their stories and their photographs. I am just the one who has had the enviable task of putting it all together.

I am sure that after the book is published there will be more Asian Transport and Astran employees who will wonder why they have not been featured. As hard as I have tried, it has not been possible to locate everyone and on more than one occasion, just when I thought I had found someone else, the trail has suddenly gone cold, or in some cases I have made contact, but sadly I have not been able to persuade the person to contribute anything.

There were also many sub-contractors whom I have not found or been able to feature, so for all of you please accept my sincere apologies for not including you.

Last, but not least, thanks to my very good friends Paul Davis and Marcus Lester who have given me their valuable time to help with the cause. Thanks also to Peter White who has proof-read and to the very talented Chris Stringer who has designed the wonderful book cover and also made a great job with some graphics and editing some of the images.

Finally, I hope I have made a fair job of documenting the history of Astran and that the stories and photographs here illustrate just how the men persevered and got through against all the odds, helping to make the company such a legend on the Middle East overland run. I hope you enjoy reading the book as much as I have enjoyed researching and writing it.

Ashley Coghill, 2010

Image credits

AC	Ashley Coghill	GP	Gordon Pearce	PS	Pete Sumpter
AM	Alison Moss	GS	Gavin Smith	RB	Ron Bell
ASTC	Astran collection	GW	Graham Wainwright	RC	Ron Chapman
BB	Barrie Barnes	JB	John Bruce	RE	Rick Ellis
BJ	Brian Jennings	JBO	Joe Bowser	REV	*Reveille* magazine
BL	British Leyland	JC	John Cooper	RG	Richard Garn
BP	Bob Paul	JCO	Jez Coulson	RH	Roger Haywood
BPA	Bun Parlane	JD	John Dyson	RHA	Robert Hackford
BPO	Bob Poggiani	JDE	Jan Dekker	RR	Roger Rabbit
BR	Bill Robins	JF	John Frost	RS	Ray Scutts
BV	Bobby Vallas	JH	John Harper	RW	Rob Warren
CAR	*Carpet Review*	JL	John Lewis	RWIL	Roger Williams
CH	Chris Hooper	JW	John Williams	SL	Steve Lynch
CM	*Commercial Motor*	KB	Kim Belcher	SN	Simon Normanton
CN	Charlie Norton	KL	Kevin Letham	SP	Steve Pooley
CS	Chris Stringer	KS	Kenny Searle	TM	Trevor Marks
DC	David Cassell	MS	Mark Stewart	TS	Tony Soameson
DP	Dave Poulton	MSO	Martin Sortwell	TSC	Trevor Stringer collection
DS	David Scarfe	MT	Mike Taylor	TT	Terry Tott
FH	Frank Hook	MW	Michael Woodman	TWN	*Truck World News*
FT	*Financial Times*	NB	Nick Bull	UM	Uncle Malcolm
GB	Gordon Benn	NH	Nigel Harness	UNK	Unknown contributor
GF	Geoff Frost	PC	Peter Cannon	YT	York Trailers
GM	Guy Motors	PJD	Peter Davies		
GOO	*Google Maps*	PK	Peter Keen		

Introduction

To many people, the phrase 'Continental long-distance trucker' conjures up romantic images of big shiny rigs cruising along sunny open highways.

Today's trucker is King of the Road and his ultra-modern vehicle is kitted out with everything he could ever wish for. He will undoubtedly have a powerful engine of 500–600 bhp and his luxurious cab will include a sumptuous bunk, air conditioning, central heating for the winter, satellite communications and TV, fridge-freezer, microwave, the proverbial kitchen sink and sometimes even a shower. Everything is provided to make life as easy as possible and a real home from home for him as he travels around Europe and beyond for weeks at a time.

During the 1970s and '80s, the image of the long-distance trucker was glamorised in TV shows such as the American series BJ and the Bear and the famous film Convoy which attracted an audience of millions. British TV has also featured truckers in many programmes including *The Brothers* and the hit comedy series *Only Fools and Horses*. The glamorous image continues to this day with the very popular *Ice Road Truckers* series.

Over the years there has also been much written about continental truckers with most transport publications carrying glossy features such as 'Long Distance Diary'. The world of Formula 1 cars, power boating and rock tours have all been covered in this way and have often illustrated truckers rubbing shoulders with the rich and famous.

So it would seem that the life of a modern-day long-

Today's trucker is king of the road. Two of Astran's finest on the Gulf coast 5,000 miles from home. (KL)

distance trucker isn't a bad one at all. What a great way to see and explore different countries and cities, sample exciting new food and drink, gain a taste of local cultures and get a nice tan along the way. Oh, and get well paid for it too.

But now take yourself back to the 1950s and '60s. Things were a whole lot different with just a handful of British transport companies engaged on work to France, Germany and the Benelux countries. The meat market in central Paris was a typical destination with fresh beef and lamb transported there from Ireland. That journey involved two long ferry crossings and had a total distance of some 700 miles which in those days was indeed long distance. The trucks were very heavily laden which slowed them down drastically, especially as they crawled up the infamous banks of Beatock in Scotland and Shap in northern England where top speeds were no more than 10 mph.

Living quarters in the latest rigs include everything a driver could ever wish for. There's even a fully fitted kitchen in this Scania 'Longline'. (TWN)

Vehicle makes such as AEC, Atkinson, Foden and ERF, many of which were built with wooden-framed cabs, were the mainstay for hauliers but the vehicles were so primitive that drivers would have to lay blankets over the engine hump in the cab in an attempt to deaden the deafening noise coming from beneath them. Power was limited to around 110-120 bhp and there were certainly no luxuries such as power steering, heaters or radios. Most trucks were very draughty as the wind whipped in through gaping holes around the pedals and doors.

With no sleeper cabs, the men would usually book into digs along the route or if they were running late and the B&Bs were already full they would simply sleep as best they could, propped up in their cabs. If they were lucky, they'd have a wooden board which they would lay across the seats to lie on.

During the cold winter nights engines were left running with a brick wedged on the accelerator pedal which would keep the revs up and enable the heater to blow lukewarm air, if they were lucky enough to have a heater, but as the cabs rattled all night long and filled with toxic exhaust fumes, the poor drivers got hardly any sleep at all. Trying to motivate men to do trips from London to Edinburgh was bad enough, so imagine trying to get them to do a trip to the Continent! When word got around in the early 1960s that two men had decided to tackle a new pioneering continental journey, many people in the road transport industry simply laughed at their plans. The fact that the men intended to do it in a truck very similar to the ones described here and which was for all intents only fit for local work, made it even more amusing. But the main point was that the men were not remotely connected to the transport world. Instead, they were friends who were practising and studying medicine in London.

Against all the experts' better judgment and advice, the two entrepreneurs set off for their chosen destination – Kabul, capital of Afghanistan! As expected, the journey was not without problems, but the pair battled through and made it there and back, putting a whole new meaning to the phrase 'long distance'. In fact they had undertaken a colossal challenge. Unknowingly, they had opened up the very first long haul overland commercial route across Europe and deep into the Middle East. In so doing, they were solely responsible for starting the biggest phenomenon in the road haulage industry: 'the Middle East run'.

Introduction

The mainstays of British haulage in the 1960s were low-powered, noisy trucks with wooden-framed cabs. (PJD)

This book is about the company that one of those men founded, from its roots in the 1960s as the fledgling Asian Transport, to the thriving and very healthy freight forwarding business of today that is Astran Cargo Services Limited. The company can boast at being involved in overland transport to the Middle East for over 45 years and can be extremely proud of its heritage and achievements.

The book documents the complete history of the company to date, but concentrates predominantly on the early years. For the employees of Asian Transport who blazed the trail to faraway places that other truckers had not even heard of, they would consider a journey from London to Scotland to be a mere spin which would fade into insignificance as they battled against the elements to stay alive while stranded 6,000 feet up on a mountain pass in temperatures as low as minus forty! They were the same men who would think nothing of taking on a 10,000 mile round trip to Iran or the Arabian Gulf summer or winter, 365 days a year. To them, it was all in a day's work, but to me they are my heroes and have without doubt gained themselves the very prestigious accolade of The Long Haul Pioneers.

Mr Michael Woodman, founder and owner of Asian Transport - Astran. (ASTC)

Peter Cannon, driving AMY 147H, deals with a Turkish 'diversion' en route to Iran in 1970. (PC)

*Modern-day Astran can be very proud of its momentous achievements as leaders
in overland transport to the Middle East for over 45 years.* (KL)

CHAPTER 1

The Main Attraction

So what exactly was it that attracted men to drive thousands of miles across two continents in a truck which was in truth, little more than a tin can with an engine on wheels?

In the early 1970s when newly opened routes to the Middle East were being explored and when trade relations between Britain and countries such as Iran, Afghanistan and Pakistan were at a high, getting on the 'overland run' was seen to be the ultimate driving job. Then later on into the 1980s when the construction and oil boom in the southern Middle Eastern Gulf states was in full swing, every man and his dog wanted to try their luck at the job. They had heard that it would reward them handsomely with pots of gold and they all wanted some!

The Middle East run conjured up mystical and romantic images for those who yearned to have a go. For those who did it, it was the crème de la crème, something that could be talked about for years to come. It was a true phenomenon which to this day clearly evokes many vivid memories for those who drove there. It is a subject which commands the total attention of everyone who has ever listened to an old hand telling stories of his days on the long and sometimes very lonely road to the East.

Ray Scutts, one of the first men to drive for Asian Transport in the 1960s, was looking for a challenge: 'I was always looking east wondering what was out there. There were two imaginary pages at the back of my AA road map book and I wanted to see them.'

The overland run was a job that definitely sorted the men from the boys and the very first trip was all it took to determine whether the man was cut out for it or not. To some it was a clear case of never again and they would literally dump the truck at the first sign of trouble and run home, but to others it was something

The Middle East run conjured up mystical and exciting images. (ASTC)

The long and dusty road sorted the men from the boys (ASTC).

that they had to keep doing, time and time again. They got the bug and like all jobs that people enjoy, it became habit-forming.

Two important qualities that anyone needed to do the job were self-reliance and 'bottle'. As David Miller, who worked as a sub-contractor for Asian Transport in the early 1970s, told me, 'If you weren't confident in yourself you probably wouldn't have started doing the job in the first place, but nerve was the more difficult quality to access. There were great big strong blokes who would bottle it as soon as the going got tough and little guys who had it by the crate full.'

Gavin Smith spent the best part of ten years working for Astran. Here are his comments:

'There are three types of truck driver. Those who are content to work in the UK and get home to their families most nights, those who want a bit more freedom and spend a couple of weeks driving around Europe and those who do the Middle East run. Each one possesses his own special skills but the Middle East man has all of them and more and is looked upon by everyone else as the real hardcore driver and the hero. It takes a special kind of person to drive those vast distances, someone who is confident and comfortable with himself and doesn't mind being alone for very long periods.

'The time I spent taking loads to Doha and other places in Arabia was the most challenging that I have ever spent on the road. They were the best years of my working career and I am sure that is true for the others who did it. Once you get it into your blood it stays with you forever. When I finished with Middle East transport, there was nothing left for me to drive for and even though I have now lived my dreams and achieved my goals, I still miss those golden years on the dusty roads.'

It was a very difficult, strenuous and sometimes dangerous job, definitely not your normal nine-to-five. Once the men had crossed the English Channel they were on their own and literally had to fend for themselves. A typical Middle East run would clock up a round trip of some 10,000 miles, more than some people travel in a year, and take approximately four or five weeks to complete - IF there were no hold-ups.

In the early days of the 1960s and '70s, communication back to base and home was very primitive and limited to once a week or so. In the worst scenarios such as hold-ups by bad weather, border problems, breakdowns or accidents, it was not uncommon for relatives to be in the dark for two or three weeks and in some cases much longer. As far as wives and girlfriends were concerned, they had no idea where their partners were until they got that precious phone call.

Gordon Pearce, who started with Asian Transport in 1966, wrote letters to his wife once a week and posted them home as he went along. He remembers trying to make an important telephone call to his boss from a Turkish Post Telegram and Telephone (PTT) office, the equivalent to a Post Office in the UK. 'The connection alone,' Gordon said, 'took over one and a half hours as there were so many other people waiting for a free line. When I finally got through, the line was terrible and I was only able to speak briefly. It was like being underwater.'

The normal communication method was via a telegram service but there was always the chance that it would not arrive at its destination, so using a telephone was a definite way of getting through, even if it did take an eternity. A telegram example that Pearce remembers was when he arrived in Tehran and had to let his office know his whereabouts. There was a set price for a set amount of words and he had written 'Arrived safely in Tehran. Gordon'. The telegram operator noticed that he had one more word available for his money, so he changed the message slightly to read: 'Arrived safely in Tehran. Love Gordon'.

After being out of touch with civilisation for so long, it was a relief for those waiting at home when news came through that a driver was okay. Bun Parlane's widow, Norma, clearly remembers her feelings when her late husband was driving for Asian Transport in the 1970s:

'It was an adventure for the men. Every time they left home on another trip we just didn't know if they were coming back or not. We never knew when we would hear from them next or even see them next. We would get a phone call from Kuwait to say they were on their way home, but we knew that they had some dangerous places to drive through before they got here. That was the worrying bit but we just used to get on with things as best we could. It was a case of having to.'

Even though communications improved dramatically through the 1970s and '80s, the job was still fraught with problems and danger, so those waiting at home still had concerns. June Bruce, whose husband John drove for Astran during the 1980s looked forward to the usual ten days leave that John got between each trip. Married for forty years, she has been with him through thick and thin. She explained:

'There was only so much I could cope with while John was away, so I used to have a list of the jobs that had to be done when he got home which I pinned to the fridge. I had two children to bring up and found it difficult doing all the other things as well. The ten days was nice, but by the time I'd got used to having a husband again, he was off. The first week of him being gone again was absolutely horrible, but I had some close friends who I saw and it wasn't long before my life got back to how it normally was.'

'It was easy for us,' John responded. 'We used to shed all the social responsibilities when we got on the ferries. On the way out, our last phone call home was from Adana in Turkey and then we didn't know how long it would be until we could call again.' Then June continued:

'You do worry about them a lot, but all you could do was wait for the phone calls. I remember John phoning once to say he was going on a particular ferry, then I heard on the news that it had sunk. It was utterly frantic in the house until I got another call from John saying he had missed his place on the boat and he was okay. I wouldn't say it was the best of lives, it's only the fact that we love each other so much that has kept us together.'

Both John and June are now retired and enjoying life to the full. June has the final word: 'I'm glad he doesn't drive those great distances any more. The furthest I let him go now is to our caravan in the south of France - although I do let him stop at all the Les Routiers truckstops on the way down.'

For most of the men, it was the sense of adventure with a capital A which attracted them to the job. They were fed up with monotonous and boring runs around the UK and yearned for something different. Speaking to others already on the run, they learnt of the dangers and perils of the job and, far from frightening them off, it gave them the urge to try it for themselves.

The worst part of any journey to the Middle East was undoubtedly the very long drive through Turkey. With an end-to-end distance of around 1,500 miles, the country was fraught with danger and hostility, and in the early days it would take the men between four and five days to get through.

The weather became atrocious in the winter months, especially in the east of the country where the worst

If diesel fuel became waxy, drivers had no choice but to light fires underneath their trucks to thaw things out. (GP)

Trucks regularly slid off the roads in harsh winter conditions. (GP)

mountain ranges were to be found. Within 170 miles there were three tricky passes to negotiate, each at over 6,000 feet above sea level. If it wasn't the diesel fuel waxing up and becoming too thick to flow through the fuel filters, it was the blizzards which were an everyday occurrence and caused the narrow roads to be blocked, halting trucks for days upon end.

It became a constant battle against these harsh elements to keep the trucks moving. If they did grind to a halt drivers had no choice but to light fires under the fuel tanks and engines to try and thaw out and warm up the waxy fuel. If they were unlucky enough to lose their trucks off the road, they would have to dig them out with a pick and shovel.

At night temperatures in the formidable Turkish mountains regularly plummeted to minus 30 degrees centigrade and sometimes minus 40. Gordon Pearce remembers being stuck for five days in such conditions:

'The engine was frozen solid so at night I used to go to bed completely dressed in three pairs of socks, two or three pairs of trousers, a vest, shirts, jumpers and a coat. I also wore a woolly hat. I got into my sleeping bag and laid three or four blankets over the top. Inside with me, I cuddled my small gas bottle to keep it warm so it would at least work in the morning and heat water for a cup of tea. I also had a thermos flask which I'd filled with water earlier in the day.

'I couldn't see anything out of the windows as they were covered in thick ice. I used to warm two pennies in my hands and then stick them on the window. When they thawed two little holes in the ice I would put my eyes up to them to see out. I've never known cold like that. You really had to look after yourself and take all the right precautions otherwise you would die. It was as simple as that.'

Running out of food and water during the winter months was another hazard with which the men had to contend. Ray Scutts told me that although he and his colleagues all took plenty of provisions, on more than one occasion they completely ran out:

'I used to heat up snow and even though it wasn't the ideal solution at least I had something hot to drink. We all used to carry Bovril which we added to the water. It was so tasty and was the main thing we lived on when we were stranded and had run out of everything else. In all the times I got stuck I never feared for my life because I always knew I would be found. It was just a question of when. Although we used to get extremely cold and we all lost weight, all of us got through in the end.'

Great care was needed to negotiate roads throughout the Middle East but especially in Turkey during the most hazardous times of winter and springtime on the mountain passes when Turkish Tonkas and Kamikaze buses were coming the other way.

'Tonka' is a word unique to Turkey. It was a nickname given to locally built four-wheeler trucks which were precariously modified with extra axles and bigger bodies to carry far more cargo and weight than that for which they were intended.

Drivers took plenty of provisions with them but had to rely on the very bare minimum during hard times. (PK)

A typical Turkish 'Tonka'. More often than not grossly overloaded and definitely overweight! (RH)

The dangerously overloaded trucks were the backbone of Turkish transport and always seemed to be driven by men for whom the word fear had no meaning. In their minds everything was fine as they usually had a tiny copy of the Koran hanging in the cab which they believed would protect them and keep them safe if anything untoward was going to happen. To avoid collisions with the Tonkas it was usually the truck drivers who would have to move over, meaning they would be the ones ending up in the ditch!

Kamikaze bus drivers were just as bad as the Tonka drivers and they would stop for nothing, even with a full complement of passengers. Found mainly in Turkey, but also throughout the Middle East, many of the buses were grossly overloaded by at least 100 per cent and they would be driven flat out, with no regard for anyone or anything. All buses had a motto painted on the front which read 'Mushalla', in praise of Allah, and the philosophy was that if a crash was going happen, then so be it, Allah will look after us.

Those who chose to do this Middle East run had to be very tough, resilient and dedicated. To them the word survival was the keyword and they knew it intimately. They were not just ordinary truck drivers. Instead, as Eddie Parlane told me, 'We were idiots.' Eddie, who worked for Asian Transport in 1968, went on to declare, 'Who else would do that type of job, getting stuck halfway up a mountain, putting on bloody snow chains in the freezing cold, lighting fires under the truck and being stuck up to your neck in mud, all for twenty quid a week? Our wives didn't like the job we did, but it was something we simply had to do.'

It would seem that a sense of adventure was definitely in the blood of these extraordinary men. Bob Paul who was transport director at Asian Transport and latterly Astran, and who has probably more experience of driving to faraway places than anyone, sums it all up in one simple sentence: 'In the winter months, if you said that your guts weren't in knots and you weren't petrified, you were a liar.'

The Main Attraction

Kamikaze buses were a notorious hazard. Most were grossly overloaded like this one in Syria which was doing over 60 mph! (MS)

If the bitter cold temperatures of the Turkish winters weren't enough to cope with, the men then had to endure the extreme contrast of the vast open deserts of Arabia. The searing desert heat has even been known to kill men when they have stepped out from their cool air-conditioned truck cabs.

Peter Fanning was working for Astran during the 1980s when exactly that happened to him. Bob Paul explained:

'I used to always, always warn my drivers to be very careful with the air conditioning especially when they got out of their lovely cool cabs as the sun would hit them instantly. Peter was at Doha in the height of summer and had spent the night in his cool cab. The next morning he got up, jumped out into the intense heat and drank three or four pints of Coca Cola. Within minutes he'd had a heart attack and died.'

When the first drivers ventured towards Saudi Arabia, there were no proper roads across the deserts, just sand tracks which had been previously made by local trucks. Good progress had to be made to cross the vast distances as quickly as possible so as not to become stranded in the heat, but it was still necessary to spend occasional nights out in the open as the lengthy journey could not always be completed in one day.

Totally aware of the hazards, Gordon Pearce was very fond of this stage of his journeys. With no recognisable signposts, he taught himself to use the stars for guidance at night and the sun in the daytime to navigate his truck across the wilderness. At night, he parked up and slept on top of his trailer with just clear skies and stars for company.

As Pearce criss-crossed the deserts, he was very fortunate to make friends with an Eastern prince and some Bedouin tribesmen. The friendly Bedouin invited him to spend one evening sharing a freshly slain goat for dinner with them. This was very tasty but a big cultural shock as he watched it being killed, prepared and then cooked.

If it wasn't the elements that held men and machines up, it was the officious bureaucrats along the routes who were out to make a fast buck and were known to change the rules without warning to suit themselves. Drivers had to keep up to date with the multitude of documents and papers they had to handle on any one trip. That was a challenge in itself and if documents were not word perfect, it would normally hold them up for a very long time. Border procedures and formalities kept changing every few weeks so the men needed to be vigilant and would have to rely on others who were coming the other way for advice.

No two trips were ever the same and there was never a dull moment which certainly made the job very interesting. Things were made more difficult as drivers had to learn how to fill in many different forms which needed stamping and processing at each border before they were allowed to carry on with their journey. The German system was particularly difficult to learn and any good driver would need quite a few trips under his belt before he understood it fully. Speaking the local language was also very useful and many of the men learnt to converse with basic words. They became experts, using bits of four or five different languages and gestures to string a sentence together. The officials became used to them doing this so that it became a system which worked well.

Gordon Pearce remembers an incident at Kapikule, on the Bulgarian/Turkish border, when he was asked by an official to describe the goods he was carrying. The official used a dictionary to translate Pearce's paperwork word by word into Turkish but he did not fully understand some of the words on the manifest. He asked Pearce what 'taps' were. Pearce made a gesture with his hand as if turning on a water tap. The official understood then asked what 'dies' were. Again Pearce gestured, this time at his shirt and then his trousers, then his jacket, pointing to different parts. Eventually the official understood he meant the colours in the fabrics.

Luckily for Pearce he continued to his destination without any further enquiries as to what was in his load. Had any more officials stopped him en route and requested to examine the goods, they would have found that in fact Pearce had accidently described the wrong types of taps and dies. He should have described machine tooling parts. It never occurred to him at the time as he could not see what was in any of the crates he was carrying and just presumed the goods were indeed what he described. Had he been stopped, he may well have landed himself in an awful lot of trouble for such an innocent mistake.

It has been known for some drivers to be thrown in jail accused of fraud or smuggling. This happened to two Swiss drivers who were employed as sub-contractors to Astran during the 1980s. Their paperwork indicated they were carrying nuts and bolts and when they arrived at Kapikule, the customs officials requested to examine the load. One of the crates was inspected but instead of nuts and bolts, the officials found large amounts of white powder! The chief customs officer was summoned to sample the mystery powder and as soon as he tasted it he presumed it was cocaine.

Both drivers were immediately hauled to jail where they were kept for over a month without being told what was happening. They endured a terrible time and were even shackled with chains. Eventually they were accused of carrying drugs, a very serious offence which took a lengthy period of time to resolve. Bob Paul flew out to Turkey on numerous occasions. With the assistance of the British consulate representative and through the Turkish transport association, he found a lawyer who eventually secured the release of the drivers. They were both found innocent at their court appearance.

The problem had been created when the documents were written out by a clerk in the exporter's office before the truck departed on its journey from the UK. Unfortunately, there were some unidentified goods in the load and so they were described simply as 'nuts and bolts'. The white powder was totally harmless and was actually a mixing compound used in the building industry. Amazingly, one of the Swiss drivers was not put off one iota by his ordeal in the Turkish jail and continued working for Astran for another year or so.

The further east the men travelled, the worse the border controls got and each country seemed to have its own unique system of dealing with trucks as they entered the particular country, which inevitably held the drivers up, in some cases for very long periods.

Dave Poulton explains the hassle he endured at the border post of Bazargan, entering Iran from Turkey, as he made his way to Pakistan:

'The queue was 121 kilometres long and took me eight days to get to the border from the back. There were three lanes of traffic for the whole length of the queue. There were thousands and thousands of trucks and the congestion was just unbelievable. And when I got to the end it got even worse with everyone pushing and shoving. Mirrors got broken, fights broke out and there was just one little archway that we all had to squeeze through. Quite unbelievable really.'

During the 1980s, Iraq had a unique formality regarding tyres, each of which was, and still is to this day, embossed with a unique serial number. When he arrived at the Iraqi entry border from Turkey it was the driver's responsibility to log every one from his truck and trailer, which sometimes totalled more than twenty, onto the relevant paperwork for the officials to check. This was a very time-consuming and awkward process which led to severe delays similar to the one described above by Poulton. To make the process even more difficult, the driver had to crawl underneath his rig to look at the inner tyres, not a pleasant task in the depths of winter.

Once the driver had got to the other end of the country, the serial numbers would have to be methodically checked again by the officials at the Iraqi exit border, to make sure they were the same tyres that entered the country.

Before this system was set up, the Turks were entering Iraq with brand-new tyres on their vehicles, then selling them at over-inflated (pardon the pun) prices and buying very cheap, worn-out ones to return to Turkey. They were making huge amounts of money, so the Iraqi government put the tyre number system in place to stop it.

The Jordanian/Saudi border at Haditha was also renowned for some terrible hold-ups. Gavin Smith, who had his fair share of them, explained the system:

'After entering the huge, dusty compound and completing passport control, the trucks were then presented for "treftish" (inspection). An Indian worker known as a "jingly", employed for labouring duties, would follow a customs official around the truck. At various points the official would mark a point somewhere on my rig with a piece of chalk and the jingly would open up that particular area. How did he open it? He used either an angle grinder or an oxyacetylene cutting torch!

'The official wanted to see inside trailer boxes, diesel tanks and sometimes inside parts of the cab itself. The small section below the windscreen and above the radiator was a favourite place to be searched as all these areas were hollow and could easily have been used for smuggling various items into Saudi Arabia. Those of us who did the overland run on a regular basis used to drop off our trailer belly fuel tanks and tool boxes in southern Turkey to avoid having them cut open and rendered useless. We would also have large hatches installed in our long-range fuel tanks on the trucks so the officials could look in without cutting any more holes. An inspection would normally take up a whole morning and afterwards we would have to put the damage right by whatever means we could. Normally a hammer was enough, but sometimes we would need welding doing which obviously cost money.'

After treftish, drivers would enter another compound where the load manifests were submitted into the customs office. There was only one small window where all the drivers had to pass their documents. It was a ridiculous set-up and sometimes there would be dozens of drivers all pushing and shoving one another to get their papers in. It was literally the strongest man who got there first and it wasn't unusual for fights to break out.

All the paperwork had to have been previously translated into Arabic – no easy task - and if any parts were wrong, the officials would simply throw the offending papers out of the window and across the room for the driver to find and somehow put back in order. He would then need to seek out a local agent and pay him to rectify the mistakes. Then the poor man would return to the customs office to try his luck again in the queue. This took a lot of precious time and it has been known for drivers to be stranded for three or four weeks while waiting for documents to be corrected. Smith continued:

'I used to show my face at the window and look for a Customs man I knew. Sometimes I would be lucky and get called inside the office where I was invited to sit and drink chai tea and eat watermelon with the customs chief while my documents were processed. This often saved me two or three days. I think they liked the English and it definitely helped that I could speak a good amount of Arabic.'

Patience was high on the list of personal qualities that a true Middle East driver needed. John Bruce clearly remembers being held up at the border into Muscat for nineteen days due to a simple passport detail. He had already travelled over 5,000 miles from the UK with no hold-ups at all at any previous borders.

He explained: 'My name is John Leonard Bruce and that is what it said on my paperwork, but my passport said John Bruce so they wouldn't let me in, it was as simple as that. I had to grin and bear it and just sit there while the authorities sorted it out.'

Gordon Pearce also remembers being held up quite a few times:

'I was stuck for three or four days at a border. I had an agent make out all the necessary documents for the customs officials to stamp and verify, but it wasn't just mine that the agent was dealing with. When I saw him, he had a whole pile of folders for many other drivers who were all waiting alongside me. I kept my eye on the customs men for a good few hours as they slowly waded through the pile and I got quite hopeful when I noticed my folder was near the top. Just when I thought it would only be a short time before I was processed and allowed to go, another agent came along and dumped his pile of folders on top of ours. That was it; all I could do was step back and wait for the pile to decrease all over again and I knew that would take another day or so. The job certainly tested your patience.'

Other drivers preferred the baksheesh, or back-hand, approach to bribe the customs men to process their paperwork more quickly than the others in the queue. British drivers were liked by the officials as they always had Western goods with them. Usually a packet of cigarettes, a box of coffee or tea, or even a cheap pair of jeans would do the trick - but by far the best currency was a pornographic book which had been bought in Holland or Germany on the way through. As the documents were handed in for customs scrutiny, there would be a bulge in the middle of the folder indicating that a gift was present.

Sometimes fines were imposed but receipts were never issued. One driver was fined 200 Turkish lira for overstaying his visa time and not exiting the country before the expiry date. The fact that he had been severely held up in a queue of 300 other trucks was of no relevance to the officials.

Queuing was very much part of the job. Tony Soameson and two other Astran drivers wait patiently in line at Zahko on the Turkey-Iraq border. (TS)

The camaraderie was unique on the Middle East run. Ray Scutts, driving UHM 25F, helps a stricken Dutchman by towing him up a particularly nasty track in Turkey. (RS)

Ray Scutts had a tried and tested method of overcoming some of the frustrations he suffered at the hands of the officials. He told me, 'I used to get right in their faces and grin at them like a Cheshire cat, jump up and down and wave my arms and hands about and shout, "Me crazy Englishman. ... Up Manchester United!" Normally it worked and they let us through.'

Of course the job wasn't all doom and gloom. The one thing that kept drivers going and sustained their morale was the camaraderie amongst them which was unique on the overland run. Without doubt, every man who drove to the Middle East knew that wherever he was, or whatever situation he was in, there would always be another driver who would stop for him, just to check that things were okay. Whether he was parked up and resting, stranded at a border, broken down or involved in an accident it wouldn't have been too long until someone passed by and that driver may not have even been from the same country. The Bulgarians and Hungarians were particularly renowned for their friendliness and helpfulness in the early pioneering days.

After driving alone for days at a time, it was a welcome sight and a big relief to see other trucks parked up together along the route. There were many famous truckstops which drivers would head for at a particular time so as to keep on schedule, or to meet up with work mates who they hadn't seen for a while. There was always much merriment as the men ate, drank and talked together and they would think nothing of spending two or three days there having parties and catching up.

Two such places were in Istanbul, Turkey - the Mocamp and the Londra - which were quite close to each other. Both truckstops provided everything the drivers needed and they were used as regular rendezvous points, or in the case of some less fortunate drivers, they were safe places to stay if the men had lost their trucks or run out of money. And for those who had a few days spare and fancied a spot of rest and relaxation, there was a motel, a night club and even a swimming pool on site.

Even though they enjoyed the socialising aspect of the job, most drivers on the Middle East run would class

John Frost (far right) meets up with some old mates from Hungary's state-owned transport company, Hungarocamion. (JF)

themselves as loners who preferred their own company and would much rather be back out on the open road. It was no job for anyone who got homesick or who wanted to spend time with their family. Living for weeks on end in a truck was no good to anyone who suffered from claustrophobia either. One cannot think of any other job which would require eating, sleeping, washing and working in the same 8-feet-square room for weeks at a time.

Being independent went hand-in-hand with the lonely lifestyle and many drivers would set their own pace to get the job done to the best of their ability. Once they had mastered the routes and borders, they quickly learnt that while one journey may take ten days, the next identical one may take twenty or thirty as they never knew what hold-ups were round the next corner.

Dave Poulton was never one for running with others and much preferred to go it alone, stopping and starting when it suited him. He was featured with three other Astran drivers in the BBC TV documentary *Destination Doha.* When asked about his job, he said on film that, 'Most days I start driving at five or six o'clock in the morning, drive maybe four hours, stop for a cup of coffee and maybe a sandwich. Then I carry on until the evening when I have a good cooked meal.'

In today's world of time-sensitive deliveries and logistics, everything is arranged to a very tight schedule. Many of the old hands could never work to these schedules and they laugh out loud at the thought of today's trucks being booked into RDCs (regional distribution centres) at ridiculous fifteen-minute intervals. Ray Scutts summed it up: 'If you are even thirty minutes late nowadays they refuse the load and tell you to piss off. Imagine travelling 5,000 miles and being told that. We never worried about deadlines. We got there as best we could.' A laid-back attitude was certainly a major attribute that the men needed to cope with the pressures and hassle they endured.

Another attraction to the Middle East run was the urge to travel. Many of the men were from military

*The Londra truckstop in Istanbul, Turkey in the mid-1970s.
This was a regular place for drivers to take some well-earned rest and catch up with mates.* (RWIL)

backgrounds or the merchant navy and were well used to moving around and had a good understanding of different countries and cultures. Some had been travelling in their youth and found the job to be the perfect way to continue their nomadic lifestyle.

They were envied by those drivers who were engaged on local European work and as Ray Scutts explained, some of them just couldn't imagine that such long journeys to the Middle East even existed:

'I was on the ferry once and two other drivers got talking to me. They told me they were going all the way to Paris and it was going to take them at least three days. They were quite frustrated that they would be away from home for so long. I shook my head and told them I was going all the way to Tehran and it was going to take me at least six weeks until I was home again. They both looked at me in utter disbelief. Neither of them had even heard of Tehran and couldn't believe I was going to drive my truck all the way there and all the way back.'

Just about all the men who drove the vast distances had itchy feet and no sooner had they returned home from one trip, than they were champing at the bit to go back for another one. They could not sit still for one moment and simply had to keep travelling. The further they went, the better it felt.

Take the case of Dave Poulton. He thoroughly enjoyed his time on the Middle East run managing an amazing 123 trips in the sixteen years he drove for Astran. 'For me it wasn't the money,' he said. 'It was a job I loved and as soon as I climbed in my truck, shut the door and left England it was a different way of life. It was the independence and freedom that attracted me. Once you were on the other side of the English Channel, you were on your own and that suited me perfectly.'

I have asked all the men I have interviewed this same simple question: 'Would you do it again?' Every one of them has replied with a resounding YES. As Eddie Parlane said:

John Frost surveys his domain. It was the independence and freedom which attracted men to the Middle East run. (ASTC)

'If circumstances allowed me, I would jump at the chance to do it again, but only as it was in the early days when it was a real challenge, exciting and adventurous. It was the most amazing experience and totally unlike any other truck driving job I have ever done since. It was totally unique and something that will never be repeated. The memories of my time driving to the Middle East in the late 1960s will stay with me forever.'

The men had a great life and a good laugh, although at times it was frightening and depressing. Some of their stories are beyond belief and there is no doubt in my mind that these men are from a very special breed. With grit, determination and bags of enthusiasm, the main attraction for all of them was the ultimate sense of adventure.

CHAPTER 2

In the Beginning

In the spring of 1964, Michael Woodman and Bob Paul set off together from the rolling hills of Cheshire, England, to begin a mammoth 5,000 mile overland journey which would take them through nine countries to their final destination Kabul, capital of Afghanistan.

The idea for this epic journey originated from Woodman who was a qualified dentist who had done his National Service with the British Royal Air Force dental corps in Singapore. When he was de-mobbed, he decided to drive his family back to England! They weren't allowed into Burma, so they boarded a ship to India and from there drove overland.

As they came through Afghanistan, Woodman noticed a vast amount of freight pouring into the landlocked country. He began to wonder if consignments to faraway places could be sent by the roads he was using rather than by the traditional method of packing them into shipping containers. He thought that trucks might be able to make quicker journey times and keep costs down.

Woodman also knew the hazards of sending goods in containers. Ships were at sea for many months at a time and if they encountered rough weather, the chances were that freight would get damaged or lost overboard. There was also the possibility of theft as the goods were handled many times before arriving at the final destination. He had experienced damage and theft from containers during his time in the RAF and he was not impressed. He was fully aware that this did not please customers who simply wanted to see their products arrive as economically, quickly and safely as possible.

During his student years in the late 1950s Woodman had met Bob Paul while they were both training at Guy's Hospital, London. The pair became good friends and while Woodman was away in the Far East, Paul continued his course in medicine. After returning to

The inaugural route that Woodman and Paul took to Kabul in the spring of 1964. (GOO)

the UK, Woodman took up a post at Guy's where he rekindled his friendship with Paul. The pair spent much time together socialising and playing hockey for the hospital team and Paul was even lucky enough to have gold fillings fitted by his friend.

To his colleagues all seemed well, but Paul was doubtful that he wanted to follow in his father's footsteps in a medical career. He couldn't see himself sitting behind a desk prescribing pills to people all day as he was more of an outdoor person who enjoyed life to the full. Deep down he yearned for something more adventurous.

While Paul was at a dinner party hosted by Woodman, fate played its part and made a decision for him which to this day he does not regret. Paul explains:

'We were talking about my career and whether I should carry on and finish my studying or just drop it all. Mike then told me about driving back overland from the Far East and unfolded his idea about taking a truck to Afghanistan. I was intrigued and it interested me greatly. I thought to myself, I could drink to my heart's content and wouldn't have to worry about killing any of my patients. It sounded good and it was different so the next day I told Mike I'd thought about it and my answer was that I'd give it a try.'

That was all Woodman needed to hear. He had been sleeping on his idea for a year since returning to the UK and now it was time to put things into action.

Setting up a brand-new limited company from scratch was, and still is, an expensive and time-consuming affair, so Woodman made enquiries about buying an existing business which had ceased trading. He went to Companies House in London and paid about £100 for a business named Scandsmith Road Transport Limited.

Woodman then joined the Road Haulage Association to obtain a Trans-Internationales des Routiers (TIR) carnet which he needed to be able to undertake international road journeys in a truck. Next he applied for and was granted a B licence which he had to have before he was able to run a commercial vehicle for hire and reward. Once these three obstacles were overcome he was able to look for work.

Considerable time was devoted to finding a suitable load. Woodman and Paul approached numerous exporting companies, but they were all wary of sending freight on such long overland journeys. As it had never been done before, many potential customers feared losing valuable stock, so were simply not interested. Eventually and with much perseverance they found themselves knocking on the doors of Linotype and Machinery Ltd (L&M), a company specialising in the manufacture of typesetting machines. These had become popular in the printing industry throughout the world, but especially where British colonialism was to be found. L&M were already exporting to faraway places such as Bombay, Hong Kong, Singapore, South Africa and Egypt, and now the company had an £84,000 contract on their books for Afghanistan.

Woodman and Paul put it to L&M that sending the machines by road would greatly reduce packing costs. Furthermore, the risk of damage and theft would be virtually eliminated by their new driver-accompanied service via the pioneering 5,000 mile route which they had worked out through Europe, Turkey and Iran. After much discussion they left the meeting with nothing to show for their time and effort.

Woodman and Paul thought their idea had been turned down, but a few weeks later and completely out of the blue Woodman received a phone call from L&M. There were three loads to go and he was offered the job. As Joe Swain, Shipping Manager for L&M, said in a press article entitled 'Original Shipper and Overland Carrier Reunited':

'At first it sounded like a crazy scheme but after looking at it carefully, I felt it might just work. At the time I had a contract to deliver typesetting machines to Kabul for the Afghan government and the only way to ship them was by sea via Pakistan which would have taken at least eight to ten weeks.'

Woodman's overland journey back from Singapore had given him vital experience and knowledge, and he believed that without those two very important factors, he would not have gained the initial contract. He may well have clinched the deal when he explained to L&M that he could deliver the machines much more quickly than the alternative. This was by sea to Karachi, off-loaded and put onto the railways for the journey to Peshawar and then transhipped again onto trucks for the final leg through the Khyber Pass and into Kabul. This method would have included a land journey of over a thousand miles after leaving the ship so that the total journey time from door to door would have been in excess of three months.

As founder and managing director, Woodman was now properly in business and Scandsmith Limited had its first load, but unfortunately as yet there was no truck to haul it so he had to work fast in order to keep to L&M's schedule. He took a huge gamble and invested his life savings of about £2,000 in a second-hand 1962 Guy Warrior tractor unit, 125 bhp, registration 10 RMY.

However, as his money was extremely tight Woodman could not afford to buy a trailer. Instead, he went to York Trailers and asked them to lend him one. He managed to convince them that with their name and logo painted all over it, it would do them the power of good and so in return York's agreed to Woodman's request.

There was now just one final obstacle to clear before the men could depart for Kabul. Neither Woodman nor Paul had ever driven a truck before, and as the departure date was looming, they had to learn very quickly.

As luck had it, they were drinking in their local pub and got chatting to a truck driver. After explaining their predicament to him, he agreed to give them both a lesson which they decided to combine with the collection of the trailer and the run up to the L&M factory at Altrincham near Manchester. In those days there was no such thing as an HGV (Heavy Goods Vehicle) licence, so it was no problem for the men to get behind the wheel of their truck.

Paul takes up the story:

'We were going up a bit of the M1 and the old boy was sitting on the engine hump in the middle of the little cab while Mike and I took it in turns to have a go at driving. He was showing us the gears and pointing out all the bits and bobs around the cab. We soon got the hang of it to which he said, "Forget it you two, you are naturals. You don't have to worry about anything."'

With everything shipshape, they confidently drove into the factory to load up. Documents were provided

The Guy in Corby high street after collecting the trailer from York's, April 1964. Note luggage and water barrel strapped to cab roof. (YT/CM)

by L&M and the trailer and its contents were examined by inspectors from HM Customs and Excise Department before they fitted a seal to the rear doors. Special comprehensive insurance for the vehicle and the journey had already been arranged by Lloyd's of London, so the long trek to Kabul could finally begin. It must have been an extraordinary sight to see the little truck pulling away from L&M's warehouse with luggage strapped to the roof and the words GREAT BRITAIN-AFGHANISTAN EXPRESS splashed all over the trailer.

The journey back to London was put to good use with Woodman and Paul again taking turns to practise driving while their instructor kept a beady eye on them. The three had struck up a very good friendship so it went without saying that a huge celebration was held back at the pub where they first met before they went their separate ways.

Woodman and Paul learnt to drive in the tiny cramped cab of the Guy Warrior as they went along. (PJD)

On Wednesday 29 April 1964 Woodman and Paul made their way out of London and towards the docks to begin their journey to Kabul. It was here that their first problem arose. Again Paul takes up the story:

'We were going quite well when we just rolled to a stop. Mike and I got out and had a look round to see what was wrong. Neither of us knew anything about mechanics and to make things worse we didn't have any tools. We stood there for a while scratching our heads and wondering what to do when a lorry pulled up and the driver asked if we were having trouble. I told him we'd stopped and we didn't know what was wrong so he had look himself and told us we'd run out of diesel.

'Unfortunately none of the gauges on the dashboard worked properly, so we weren't to know the problem. He had a spare can of fuel which he let us have and then told us that all we had to do was bleed it through and we'd be rolling in no time. Mike and I looked at each other for a minute bemused because we didn't understand what he meant. The only bleeding we knew about was in the medical world.

'He kindly stayed to help and showed us how to get the engine going again and then he asked where we were going, so I told him Kabul in Afghanistan. He stood there staring up at the trailer with his mouth wide open. I will always remember the classic expression on his face and his exact words were, "You're fucking mad!"'

That lorry driver may well have been correct. Woodman and Paul had no idea what lay ahead of them. They were indeed novices and their truck was extremely basic with no mechanical modifications to help it across the arduous terrain which the men would encounter. The only additions were a set a Michelin XY desert tyres which Woodman had fitted, the roof rack to hold their luggage and some water barrels.

Armed only with large-scale relief maps, a compass and a few tins of basic food supplies, Woodman and

Paul resumed their journey feeling confident that they would complete the job for L&M. As Woodman had returned overland from Singapore he had noted the various conditions along the way. In an article in *Commercial Motor* journal he said:

'I am only following the old caravan routes that the Eastern traders used centuries ago. The roads are very rough, admittedly, but one sees plenty of Russian commercial vehicles travelling from the north down to Afghanistan so I see no reason why British commercial vehicles should have any difficulty.'

After crossing on the *Gaelic Ferry* from Tilbury to Rotterdam the men found themselves trundling along the roads of mainland Europe. They were pleasantly surprised that things were indeed going so well, but unfortunately that luck was soon to run out. As Paul was driving down a mountain road near Graz in Austria he heard a terrible knocking noise. Fearing the worst but not wanting to stop in such a precarious position, he managed to keep the truck rolling to a nearby garage. The bad news was that the engine had completely given up and would go no further. This was gut-wrenching news that Woodman and Paul did not want to hear!

The garage owner, Herr Shakemeister, told them he could fit a new engine but that they would have to supply it. This was obviously going to take quite some time, so the men booked themselves into the adjoining hotel. Woodman made contact with the garage in Morden, Surrey, where he had bought the truck and explained the problem.

At first the garage wanted a lot of money, but eventually after some heated debates and threats of court cases, they agreed to supply a new engine and pay for everything. They discussed air-freighting the engine to Graz but that was far too expensive, so Woodman decided to hire a van. He and Paul drove it all the way back to Ostend, Belgium, to collect the new engine from the docks.

The trip to collect the engine went well until they got back to the border between Germany and Austria at Walzberg where they were told they had to pay an extortionate amount of import duty. Woodman and Paul tried to explain as best they could that the engine was to replace a broken one and that it would then be literally driven out of the country. None of the officials would listen or believe them and finally, in despair, Paul drove the van up to the border barrier, blocked the road and refused to move until the problem was resolved! It didn't take too long for the chief of customs to arrive.

Paul asked him for a policeman to come and witness the work being done, then he would have the evidence that the engine wasn't staying in the country. 'I think the chief got fed up with my ranting and just gave up his cause and let us go through,' Paul explained.

Relieved, the men continued back to the garage where Shakemeister got to work and replaced the broken engine. Paul and Woodman then had to keep themselves occupied while they were stuck there:

'There wasn't a lot to do but keep an eye on the work being done and try and pick up some tips on the mechanical workings. The hotel owner's father was always in the background watching and listening to everything that was going on. He couldn't speak a word of English but one day he beckoned to show us something. We followed him into a barn where his beautiful BMW convertible was kept. It was a very old one but was immaculate and he was obviously extremely proud of it. He then opened the doors and ushered Mike and me in. He took great delight in driving us for a long spin which we thoroughly enjoyed.

'The hotel was lovely too. It was room plus breakfast and that was a German word I learnt very quickly. Frühstück (breakfast) was delicious and definitely compulsory every day. There was a notice in our room saying no food should be brought in but because we were somewhat short of cash we tried not to eat out too often, so we used to buy garlic sausage, bread and so on and smuggle it all into our room. We kept the things in the cupboards and drawers and I always remember the chambermaid commenting to us once in her best English, "Ooooh, your room stinks. Vot are you doing in zerr?"'

The whole episode had cost Woodman and Paul around three weeks in lost time and at any point they could have simply given up and flown home but they never gave it a thought. They had a job to do and they were determined to get it done.

With the new engine installed and the truck up and running again, they continued through Austria and headed into Yugoslavia where they encountered some terrible road conditions.

After Yugoslavia came Bulgaria whose roads were equally as bad and tested the men's driving skills as they took it in turns to steer the truck carefully through the ruts and round the potholes. The journey became very slow and uncomfortable as the little truck struggled across the rough terrain.

As if that wasn't enough, they then came up against another frustrating delay when the official bureaucracy at the Bulgarian/Turkish border post of Kapikule cost them almost three weeks waiting time. Before they set off from the UK they had researched the trip and contacted relevant embassies for advice on transiting various countries en route. Paul relates what happened:

'We were told by the Turkish embassy in London that we would need a load manifest to allow us to drive through their country as the TIR arrangement was not accepted once we got past Bulgaria. We were told we could obtain the manifest from a local agent at the border but when we got there he wanted a fantastic amount of money as a guarantee for the goods before he would produce the document. It was something like £7,000.

'We were going nowhere until it was sorted and to make things worse, we hadn't got that amount of money with us so we contacted L&M in Altrincham and then through their agent in Istanbul we managed to set up the guarantee with Barclays Bank. It meant us having to leave the truck with the customs officials at the border and use a bus to get to Istanbul. After a few days to-ing and fro-ing, one of us decided to stay with the truck while the other stayed in Istanbul until it was all sorted. It took an eternity.'

Once they had overcome the problem, they made sure the manifest would allow them to transit through Turkey for the following two trips they had agreed to do for L&M.

The road networks since leaving Western Europe were pretty poor to say the least, but when the men arrived in Turkey their eyes were really opened to the terrible conditions. As Woodman put it in the *Commercial Motor* Journal, 19 February 1965, 'After Belgrade it was literally a track, all dirt to Istanbul.'

As they approached Istanbul, they had another spell of trouble with the truck, this time an electrical problem. Paul explains:

'The electric system kept blowing fuses and shorting out everything. The two-speed axle switch on the gearstick had packed up so we had parked up and were wondering what to do when a Rolls Royce pulled up alongside. It was the British consul who was very surprised to see a British lorry and asked us what we were doing there.

'We had a lovely chat and we explained our electrical problem to him to which he simply replied, "Ah yes. That's no problem. Get hold of Lucas, they'll fix it for you. They're electrical people, you know." Then he bid us farewell, waved, wound his window up and drove off. Mike and I just stood there in disbelief. Through trial and error and more luck than judgment, we managed to get the truck going and crawled along in the wrong gear until we eventually found a local garage to get us going properly again.'

Once they had crossed the Bosphorus in Istanbul by local ferry Woodman and Paul decided to take a northward route, following the Black Sea coastline towards Iran. Woodman is again quoted from The *Commercial Motor* journal: 'The route was really rough, asphalt for a time and then bumpy, narrow mountain roads with sharp turns and loose surfaces.'

The men managed to keep the truck going, although many times they lost traction on the gradients due to

Boarding the ferry across the Bosphorus as Woodman and Paul headed into Asia. (MW/CM)

Heading eastward through Turkey. The Guy had to be pulled out of trouble on many occasions. (MW/CM)

the low engine power and the fact the truck only had a single driven axle. Travelling over mountain passes at heights of 7,000 - 8,000 feet above sea level, they dared not fully stop the vehicle for fear of losing traction when they started off again. As it was, the truck had to be pulled out of ruts and ditches and towed up hills on many occasions. 'It was a case of being patient and waiting for a grader or a bulldozer to come along and rescue us and we just would not have got through if it weren't for the helpfulness of the locals,' Paul said.

Food was also a problem especially when they were stuck waiting for help:

'We had a very limited supply, mainly tins of fruit because we preferred to eat out. Sometimes we'd have to go two or three days without any grub. We were starving but I used to wind Woodman up and it drove him mad. I would tell him I wished I was with Caroline, my wife, at home and she'd be cooking English roast lamb and Jersey new potatoes and veg. Eventually, a couple of days later when we'd got down the road a bit more, he burst back into the cab, threw a bag of spuds at me and said, "There's your fucking new potatoes!"'

When the men passed out of Turkey and into Iran they faced more bureaucratic problems which once again caused a lengthy delay. There was a specific route through Iran for commercial vehicles which were in transit, but the officials had never seen westerners who wanted to drive a truck through their country. They didn't trust Woodman and Paul to use the official route so a customs officer was despatched to ride in the truck with them. According to Woodman, the officer had had enough after only ten miles and jumped out saying he would take a bus!

Through Iran the roads weren't any better than those in Turkey, in fact they got worse. Many of the roads were little more than unmade or very poorly maintained dirt tracks, so there was dust – clouds and clouds of it – and the best the men could in some places was to drive at five miles an hour because visibility was so poor.

Mountain passes had sheer drops on one side or the other and there was no protection at all to stop any vehicles falling over the edge. As they approached Tehran the surfacing did get better but after navigating their way

Driving across the Great Salt Desert of Iran with nothing but camels and dust for company. (MW/CM)

through the city, they soon found themselves back in the wilderness and heading into the Great Salt Desert. In the *Commercial Motor* journal Woodman said, 'There was virtually nothing for more than a hundred miles except more badly maintained dust roads.'

Finally, Woodman and Paul made it to Dogharun on the border of Iran and Afghanistan where they were staggered to find customs formalities very simple especially after all the problems they had endured previously throughout the journey. 'You just fill in a form and present it to the authorities when you reach the capital, Kabul,' said Woodman in *Commercial Motor*. That must have been the proverbial light at the end of the tunnel, the magic word Kabul. The men could almost taste the cold beer served at the bar and feel the fresh, clean water running from the hotel taps.

Bob Paul completes customs formalities at the Iran/Afghanistan border. (MW/CM)

Now in Afghanistan there were no more major hold-ups as they continued on the final leg of the journey via Herat and onward to Kabul, although the truck did suffer with horrendous braking problems. Paul describes one particular incident:

'We had no brakes at all on the trailer as they were totally knackered. We were descending a very narrow, twisty mountain track. Mike was driving and taking it extremely slowly while I was standing behind the cab on the catwalk trying to balance myself while the truck rocked this way and that. I had physically to pull on the trailer hand brake which we'd modified with a length of rope to try and slow it all down. It was a very frightening experience I can tell you.'

The men were in an area where the Russians and Americans were busy constructing a new road from Herat to Kandahar. It looked beautifully smooth and it broke their hearts to see it as they drove along the potholed track nearby. When they got to the Americans' camp, they stopped for help:

'They welcomed us in and showed us round. It was marvellous, everything was underground. It was all air conditioned but, best of all, they had ice-cold lager. We told them what the problem was and the man in charge said in his broad New York accent, "Gee, what ya reeeally need is a big ole Yanky Mack." I don't think he could believe it when he saw our little truck parked outside. They made some repairs to the brakes for us and did some other general work on the truck and trailer. Luckily it was good enough to get us into Kabul and all the way back home again.'

The last few miles of the epic journey went well. Paul felt jubilant as they rolled into Kabul:

'I said to Woodman, "Where's the bloody bar?"'

'We booked into the Kabul Hotel and had a proper wash and a few beers. But before we got there we stopped on the edge of town for a bit of a clean-up. Unfortunately I had the morning urge so I took the loo roll and went off round the back of the trailer to the bog. As I was squatting there, I looked up and noticed a Land Rover had parked near us. Word had got through that we were nearby and the ambassador's wife had come to meet us. It was all a bit embarrassing.'

Before taking a well-earned rest in the hotel, Woodman and Paul took the truck to the customs compound for clearance and to be unloaded. The officials could not believe their eyes, never imagining that the machines could have come all the way overland from such a faraway place. They were amazed to see Woodman, Paul and the little truck they had travelled

Unloading at Kabul. No forklift trucks, just men with pulleys and ropes. The Linotype machinery had arrived undamaged. (MW/CM)

so far in. The importer was also delighted and when unloading had been completed, he found only one crate to be damaged.

Woodman and Paul began searching around Kabul and were lucky to find a small garage which was able to do some more repairs on the Guy. Although the Americans had done a good job, the truck was still not running properly and as the local mechanic had worked on British vehicles before, he was able to give it a good service.

Paul then had to see about bringing a load back to the UK. He explained:

'We didn't know a soul out there so we went to the British Embassy in Kabul for some advice and saw the British ambassador but it turned out to be a waste of time. We went into his office and noticed he was on the phone. He just ignored us so we just waited around for hours but it was obvious he wanted nothing to do with us.'

With that, the men walked out. As it happened, it did them a favour. Even though the Guy had been serviced, it was still not running properly. Woodman and Paul had no mechanical knowledge, so they were unsure what they would have to do to improve engine performance if things got worse on the way home. The Guy was using much more oil than it should have done, so to preserve the engine Woodman and Paul decided to cut their losses and return to the UK empty. Doing this also meant less time wasted at border crossings. In readiness for the next trip they chose to return via Iran, Iraq and then Turkey to see if those routes were any better than the outward journey.

When they were driving near Baghdad, trouble struck once again which very nearly ended in tragedy and put paid to the whole venture. Paul explains:

'When the road surface was in need of repair, the Iraqis used to pour a mix of oil and tar all over it and then, once it had solidified, the surface became good again. Unbeknown to us, they had just repaired a section. Mike was driving and as he approached the fresh repair he saw what he thought was a lovely smooth surface and simply continued to drive without realising what it was. The truck was empty and he was able to drive at around 30 mph.

'All of a sudden there was a huge jolt and the truck began skidding and swerving left and right. Poor Mike could do nothing to keep things under control and the whole rig just left the road and we finished up jack-knifed in the desert sand. It was very lucky that neither of us was hurt, so we scrambled out of the cab to investigate what damage had occurred.

'The batteries were all smashed and the cab was quite badly damaged on the back corner where it had come right round and hit the front of the trailer. We were terribly distraught and didn't know what to do. I remember lying underneath the truck trying to see if everything was intact and then I noticed a pair of strange looking desert boots right next to me. I slid myself out and looked up to see a man standing there. "Having a spot of trouble?" he asked.

'It transpired he was an Englishman who was working for the Iraqi national oil company and was driving back to his work camp further along the road. He'd spotted us and stopped to help. He was a smashing chap and took Mike and me back with him. It was a marvellous place with prefabricated buildings which were beautifully cool but the best thing was when he took us into his house to get cleaned up his wife brought us two pints of ice-cold lager. We were miles from anywhere, filthy dirty, our morale was low and we were worried about the truck - but the sight of the beer cheered us up immensely.

'Our new best friend arranged for our truck to be recovered into his camp where the workers made some repairs and got us going again. Considering they didn't have the correct tools, they did a marvellous job for us. They fitted new batteries and knocked out the dents in the cab as best as they could. We were so grateful to have help like that. In fact we got on so well with them that on forthcoming trips through Iraq I always stopped there with a supply of British bacon which they could not get. That was my way of repaying them.'

Woodman and Paul continued homeward and made good progress through Turkey and back into Europe which Paul described as civilisation. In Austria, they stopped at a restaurant where they were able to relax and enjoy some proper food and drink. While there, Woodman managed to telephone his wife who told him of some problems at home so Paul suggested that Woodman flew back to be with her while he would continue to drive the truck home alone.

At the docks in Holland Paul bought a copy of the *Sunday Times* and was astonished to see a picture of Woodman with the headline AFGHANISTAN IN 18 DAYS. After flying home, Woodman had contacted the newspaper to report the trip.

When Paul's ferry docked at Felixstowe, Woodman was there to meet his friend who immediately asked where the newspaper had got eighteen days from, to which Woodman winked and replied 'driving time!' In fact the the round trip had taken something like three months.

Paul said that the trip was not financially successful, simply because of all the delays, breakdowns and extra costs involved. The fact that they also came home empty didn't help matters. The men had very little spare money between them upon their return, so to make the second trip more viable Woodman decided to stay at home and continue with some casual dentistry and also look around for work, while Paul did the driving. Before he set off again, the truck and trailer were taken to the AEC works in Middlesex where everything was thoroughly serviced and repaired in readiness for the next trip.

In October 1964 Paul set off on trip number two, this time sharing the cab with a Mr Renner as his co-driver. The two had previous connections at Guy's hospital so Paul thought it a good idea to try him out as he already knew him.

Due to the terribly bad winter weather conditions in eastern Turkey, Paul decided to try a different route from the first trip. He chose to go south via Syria, Iraq and into Iran because he knew the road from Abadan and Khorramshahr to Tehran was in tip-top condition.

It had recently been built especially for the Shah's wife so she could travel in style to one of her palaces, so Paul thought they could travel in style too and also save a lot of time.

As with all best-made plans, things were not quite that simple but this time there was a humorous twist in the tale, as Paul related: 'When we arrived at the Turkish border with Syria we handed our passports in for verification. A day later Renner's was handed back but there was no sign of mine.'

In those early days, there had to be diplomacy and patience when dealing with foreign officials as no one had been through the formalities before, so Paul reluctantly waited yet another day before approaching the officials again in an attempt to get his passport back. He continued:

'I walked into the office and waited around. Then somehow it dawned on me that because of my dark complexion, they must have thought I was Jewish. I told them I was a Roman Catholic and I could prove it if they wanted so I began to unzip my trousers.

'There was much arm waving, chattering and shouting but I was drawn to a clerk sitting very quietly in the corner who looked extremely ill. I knew a thing or two about medicine so I went over and found his pulse was way off the mark. He was obviously in a great deal of pain so I examined him and found a huge hernia in his groin. I told the chief official that if he didn't get the man to hospital and have him operated on immediately, he'd be dead in two days.'

Coincidently, Paul then received his passport back and the men were able to drive on through Syria. Paul finishes the story: 'When we arrived at the next border into Iraq, there was a huge queue of about five hundred people all waiting to see … the English doctor.'

The men arrived in Kabul on schedule with the second consignment of printing equipment and, after clearing customs and unloading, Paul went looking for a return load. On his inaugural trip, he had taken note of the huge carpet industry and thought he could strike a lucrative deal with some of the manufacturers. He showed them his truck and advised them how it could be packed to maximise the load space. It could then be sealed and not tampered with until it reached its destination. Eventually, after much persuading and haggling, Paul talked one of them into giving him a return load. They agreed a price and shook hands for a load which was destined for Hamburg, Germany. Paul explains what happened next:

'We got to the warehouse and I saw four or five Afghan trucks also waiting. I told the manager to load our truck but he said no, we load our trucks and take them to the border then you can load it into your truck. I told him that defeated the whole object and he had wasted our time. I was furious so I closed our trailer doors and we drove off.'

Paul wasn't prepared to waste any more time and decided to drive straight back to Tehran where he thought he had a better chance of a load. He proved himself right and managed to get a deal via an agent who was employed by Oriental Carpet Manufacturers (OCM) from London.

Until Paul introduced himself their baled carpets were normally loaded onto trains from Tehran and taken via Russia into Europe, or they were put in containers and shipped out from Iran direct to the UK which would have taken approximately ten weeks. That first return load for OCM was to be the start of a very fruitful, long relationship between them and Woodman's company.

Paul struck up a very good friendship with Brian Huffner who owned the family business and the two companies worked closely together for many years. Sending the carpets via the overland routes dramatically cut the journey time from Tehran to London to just ten days so Huffner was able to supply his customers with everything they needed and in double quick time.

Trip number three to Kabul departed in early February 1965 and must have been a breeze compared to the previous two. Renner wasn't sure if it was the job he wanted, so another driver accompanied Paul with the final consignment of presses for L&M. It was like seeing long-time friends as Paul passed through each border. He had purposely made good contacts with customs officials and agents on the previous two trips which helped immensely as he progressed towards Afghanistan.

Once the final consignment was offloaded, Paul decided to have another go with the rug manufacturers with whom he had had problems on his second trip. This time they had had obviously seen the error of their ways. Word had got round that indeed a truck could provide a safe and secure means of transporting the valuable rugs and carpets to the western world.

When Paul arrived at the factory to enquire again, a load was ready and waiting. The trailer was swiftly packed and Paul and his colleague were soon on their way to Hamburg with the very first return load direct from Afghanistan. Paul had finally managed to load his truck fully both ways which paid handsomely, boosting Woodman and Paul's bank account as well as their morale.

Had Paul or Woodman ever felt out of their depth on the first trip? Did they ever feel like giving up, or that they had bitten off more than they could chew? Paul had no hesitation:

Parked near Tehran capital of Iran waiting for a return load. 'Motor transport is rapidly taking the place of beast of burden'. (MW/CM)

'No, never. We always knew we'd get through somehow. I was used to driving cars down to see my parents in the south of France when I was a student, so I had a bit of an idea, but I was totally unprepared for Yugoslavia and Bulgaria. I couldn't believe it, there were no roads, just dirt tracks and I never knew that they alone would cause so many problems for the truck. With hindsight we didn't have the right vehicle for the job, so how it got us through I will never know.

'We had basic maps, but a lot of it was by word of mouth. We would meet Turkish and Iranian truck drivers who were all really friendly and they would tell us the best routes to use.'

Woodman and Paul quickly learnt customs and formalities as they travelled through each country. Maybe it was the reputation that the British Empire still had, but Paul recalls everyone being so helpful towards them. 'We had no problems talking with the Arabs and the whole point was that they respected us. England in those days was number one. An Englishman's word was his bond and he could be trusted.' This may well have been what enabled the rug manufacturers to trust Woodman and Paul with their goods.

Paul's graphic descriptions of the problems they encountered prove conclusively how determined and committed the two men really were. They overcame some colossal hurdles which they could never have imagined in their wildest dreams as they studied medicine at Guy's.

How did Paul's wife Caroline cope while he was away? 'No problem. She was my rock. We used to send telegrams when we could. That's all one could do in those days.'

Paul wasn't married when he and Woodman left on the first trip, but by 1965 he had tied the knot. 'It may have been just before I left on the third trip to Afghanistan when I said to Caroline "I'm off to the Middle East, can you arrange everything?" When I came back it was all fixed and she'd even bought the ring,' he chuckled.

Was Paul a glutton for punishment by taking on such incredible challenges, or was he just someone who lived life to the full, accepted things as they were and enjoyed every minute of what he was doing? He certainly wasn't sure whether he wanted to sit at a desk and prescribe pills to people all day long, he thought he wanted something more adventurous. Well he certainly made the right decision for that!

The Guy Warrior truck that Woodman spent his savings on served the men incredibly well. It covered something like 30,000 miles in the year they used it and although it had some major breakdowns and cost them an awful lot of money, it did its job proving that no matter what the weather or road conditions were, the little British-built truck made it there and back time after time.

When the business had progressed and they had sold the Guy on to buy a better truck, Paul and Woodman went looking for it, to bring it back for sentimental

The Financial Times, 10 February 1965, reports the third and final trip to Kabul. (MW/FT)

reasons and have it as a show piece at their depot. They did manage to track it down but tragically, just one week before they got to it, it had completely burnt out on the motorway and all they were able to salvage was a badly damaged cab shell. What a very sad and unfitting end for a truck with such an amazing history.

CHAPTER 3

My Family

Bob Paul

Although Michael Woodman was the founder and owner of Asian Transport/Astran, it was transport director Bob Paul who hired drivers and spent his time with them driving to and from Tehran in the early days. Spending so much time on the road, Paul forged some long-lasting friendships with his drivers. 'They were my boys,' he said. 'My family - especially when they bought the beer.'

Paul never classed himself as the boss but always thought of himself as one of the workers. He was honest and up front with all of them. In return he was totally respected and liked by everyone, which says a great deal about someone who was in control of a business for over forty years.

Originally Paul and his wife Caroline lived in central London and the pair would think nothing of inviting the boys for a meal or to stay the night before they started out on a trip the next day. There were never any airs and graces at the Pauls' house and everyone got on famously. Caroline was a lovely woman to whom all the drivers warmed. Like her husband, she too had a fantastic sense of humour and soon became very well respected by everyone. One or two drivers even baby-sat while the Pauls went out for the evening.

Socialising was all part of the job and Paul would think nothing of taking his boys to the pub for a good few beers whenever he could. He was a father figure, someone to whom they all looked up in total admiration.

He was a very fair man, but would certainly not tolerate any bad doings and he made that quite clear to his employees from the outset. On the other hand, he loved to lark about and play practical jokes and even devised an ingenious way of keeping the attention of any new employees when they were starting with the company:

'I wanted to talk to them and tell them what it was really like out there. I wanted to make sure they really did want to do the job. There was one bloke who had travelled from Norfolk to see me and I invited him to stay at my house for the night because he was starting out on the road early the next day.

Bob Paul, the original 'Long Haul Pioneer'. He never classed himself as the boss, just one of the workers. (BP)

42

'After dinner and a lot of talking, I decided to show him a film I'd made about our work. It started with one of the trucks going along a road and I saw the look on his face as the film got going. It was obvious he was already bored but knowing I was the boss he was trying to show some sort of interest. Then all of a sudden the film changed to a very explicit blue movie that I had spliced in. It stayed like that for a minute, before it reverted back to the truck. The look on his face was a picture and he became glued to the film until the very end as he didn't want to miss anything.

'I was already laughing to myself before the film changed as I knew exactly what was going to happen. It worked a treat.'

In all the years that he ran the company, Paul never needed to advertise commercially for drivers. They came simply by word of mouth and recommendation, although there were one or two pieces of publicity in the press which attracted drivers and gave the impression that they were job adverts.

Drivers were fully aware of the reputation that Asian Transport and Astran carried, so any potential employee would ring first and chat with Paul before he decided whether to invite them along for an interview. 'I was and still am, a very good judge of character and I could always work out very quickly whether I was being bullshitted or not - and believe me, I've dealt with a few bullshitters in my time,' said Paul.

Paul's responsibility for hiring drivers continued until the own-account transport operations closed in 1988 and Astran relied solely on sub-contractors. During the 1970s and into the '80s Astran's fleet of trucks grew rapidly and at its biggest in the mid-1980s the company employed 20-25 drivers. Although many of the boys stayed with Asian Transport and Astran for quite a few years, there were some who only did one or two trips.

Although there is no doubt who was the very first driver to work for Asian Transport, there is a certain amount of confusion as to when some of the others came along. I have spent many hours throwing names and dates at the boys I have managed to contact. The names are all agreed upon, but exact dates of when they joined Asian Transport/Astran and the trucks that they drove? Well that's a different story and definitely a case for Miss Marple.

Many of the boys who drove for Bob Paul in the early days are now in their '70s and their memories are somewhat faded, so what follows is hopefully a fairly

A typical Asian Transport/Scania advertisement placed in the Transport Journal, *October 1971 featuring Ray Neale (right) and Ron 'Jock' Bell. (ASTC)*

accurate account of Bob Paul's family, starting at the beginning.

This was a fledgling company which was just starting out in an area of transport which no one had ever tackled before. When Woodman and Paul decided to employ their first driver in 1966 they knew no one who had done this type of long-distance work. They did have the experience they had gained from the early Afghanistan trips and Woodman's overland journey from Singapore, so at least they had some idea of what qualities the new boy would need.

Gordon Pearce

Luckily their first employee came to them via a mutual friend so they were spared the task of searching for someone, but had they advertised in a local newspaper, they would have needed at least half a page and it would probably have read something like this:

> # WANTED TRUCK DRIVER
> ## MUST POSSESS THE FOLLOWING QUALITIES
>
> - Adventurer, comedian, mechanic, chef, diplomat, salesman, bank manager, physician and linguist all rolled into one.
> - Must also be tough, reliable, trustworthy, versatile and have the right mentality and determination to work on own initiative and solve problems along the way to get the job done trip after trip.

Many people would have been frightened off by this seemingly immense requirement just to drive a truck, but there were some men out there who were dreaming of such a job.

Enter Gordon Pearce, a quietly spoken and very astute south Londoner, one of the nicest, most charming men you could ever wish to meet. Even at the age of seventy-one he still has a great sense of humour and is forever playing practical jokes.

Pearce left school at eighteen and joined the army. After serving two and a half years in the Royal Electrical and Mechanical Engineers (REME), he tried his hand at a variety of jobs including a spell driving petrol tankers, work as a car driving instructor and even a summer season selling ice creams. He heard about the vacancy with Asian Transport while working as an insurance salesman with his brother-in-law who knew someone with connections to Bob Paul.

Pearce went to meet Paul at a local pub in October 1966 and gives the following account of the meeting:

'Bob seemed very nice and told me about the first trips with the Guy to Afghanistan. He was getting a brand-new truck to replace it and needed a driver. He told me where we would be going and I thought it sounded like a good idea to drive with him to Tehran.

'I had driven tank transporters in the army at the age of twenty and I was trained as a recovery mechanic so I thought that would come in useful. I was very interested as it would be an adventure, something completely different to anything I had ever done before. I could just see myself getting a tan as I knew the Middle East was always hot and sunny.

'We had a few more beers and talked about all sorts of things. I felt very relaxed because all my other job interviews were very formal and this was just like being in the pub with a mate. Bob asked how much I was earning on the insurance job. I told him twenty pounds a week to which he said he would pay me the same. Then he offered me the job there and then. I couldn't shake his hand quick enough.'

Paul's family was now started and new boy Pearce found himself champing at the bit to see the sun and the sand. Little did he know that his first trip to the Middle East would actually be in the depths of winter with temperatures plummeting way below zero.

Being number one driver gave Pearce a real sense of pride and achievement so that he strove to better himself at every opportunity. He and Paul struck up an incredible relationship during the following eighteen months or so while they were paired up in the new AEC rig on their regular round trips to Tehran:

'I knew Bob Paul better than his wife did. I got to know all his bad habits and the best ways to wind him up. He used to get very annoyed when he'd smoked his last cigarette.

My Family

Little did he know that I'd already nicked one or two and hidden them around the cab. I used to let him sweat for a while until I'd had enough of his moaning and his bad temper, then I'd produce a cigarette and say, "Ooh, look what I've found."'

Paul's view of Pearce completes the picture of their jokey relationship:

'*Gordon Pearce was a bloody comedian and was always playing practical jokes. You never knew what he'd do next but he always seemed to keep morale high. I remember so clearly we'd parked in the scorching desert once and Gordon had wandered off for a look around as usual. Then I heard him call me from behind the truck and ask if I wanted a shower. As I walked round to see him, he squirted me with a water pistol. We were very hot, dusty and extremely tired and I could have murdered him there and then.*

'*I always told him that he was much better than me at reversing the rig. He wasn't really, but I never let on. I told him I could never get the hang of the bloody thing so I just left it up to Gordon.*

'*Many a time we'd get stuck on a diversion and find ourselves down a dead-end road and have to reverse the whole rig out. Sometimes I'd get out and watch him back if it wasn't too far or too cold, but other times I used to get my own back on him for the cigarette trick by pretending to be asleep. I used to shut my eyes and sit listening to him cursing and swearing as he struggled to get the rig out of an awkward spot. It was actually more of a struggle for me not to laugh and why he never thought to wake me up to help me I'll never know.*

'*We used to off-load carpets behind St Paul's in London along some very narrow little lanes. The crowds would come along and clap as Gordon manoeuvred in. I must admit, his reversing was brilliant and quite a sight. It must have been all that practice on those diversions.*'

Through his charm, wit and charisma Pearce made many friends and some very good contacts during the seven years he spent at Asian Transport/Astran, but above all else he became very well respected by Woodman, Paul and all the other drivers.

Pearce's favourite truck was the AEC, but it did have one bad problem which has affected him severely to this day:

'*It was terribly noisy. You couldn't hear each other above the engine and literally had to shout to make yourself heard across the cab. I used to say everything twice. The first time to get my mate's attention and then again so he understood it. It really was that bad.*'

Gordon Pearce was Asian Transport's first driver in 1966. Here he is at the wheel of a new Scania 110 rig BGH 172H, c1970. (ASTC)

After spending so long in the AEC, Pearce began suffering with his hearing and, sadly, to this day suffers from tinnitus.

Graham Wainwright

About six months after Pearce joined Asian Transport, Graham Wainwright came along during the summer of 1967. Pearce remembers him simply by the fact that he had size fifteen boots!

Born and bred in Gloucester, Wainwright was driving for British Road Services and took his truck anywhere in the UK, sometimes staying out all week. He earned eight shillings for a night out which paid for an evening meal with bed and breakfast and left him with enough money for ten fags and a cup of tea in the morning! He was working 24-7 and even though the money was good, he yearned for something better. 'The Middle East was something different and no one had ever done it,' he said.

He managed to find a phone number for Bob Paul and after a chat attended an interview and then went for a test drive. All seemed okay and Wainwright went back to his regular job wondering if he had made a good enough impression. He heard nothing for quite a while and thought that perhaps he had been forgotten. Then out of the blue he received a phone call from Paul offering him a job and asking him to start immediately.

On his very first trip for Asian Transport, Wainwright drove a borrowed second-hand Leyland Beaver tractor unit, BLH 764B, towing a 30-foot trailer loaded with machinery. He was alone in his rig, but followed Pearce and Paul who were in another borrowed drawbar outfit, SVB 500F, all the way to Tehran and back to the UK.

'I think the round trip took about eight weeks. The old Leyland went there and back without a whimper,' Wainwright said proudly. 'I did one trip in the old girl and then had a new Scania Vabis which I shared with Bob Paul for a couple of trips.'

Cyril and Eddie Parlane

Shortly after Wainwright started, Cyril Parlane joined the family. Known to everyone as 'Bun', he sadly died in February 2008 at the age of seventy-seven so precise details of his employment are somewhat sketchy.

Graham Wainwright (centre) on his inaugural trip flanked by Bob Paul (left) and Gordon Pearce. (GW)

However, his widow, Norma, clearly remembers him sharing trucks with Pearce and Wainwright on his first few trips while they taught him the ropes.

Bun's younger brother Eddie Parlane also worked for Asian Transport for a time. Eddie was previously driving buses for Greenline and had only ever driven trucks while he was in the army. He had been to Germany and Jordan where he had some limited experience of desert driving. Brother Bun told him of a vacancy and recommended him to Bob Paul.

Eddie's first trip was in December 1967 when he went with Pearce to Tehran in the AEC:

'It was a very good introduction. I found Austria fascinating. As we drove through on New Year's Eve, the road was very narrow and close to the lovely chateaux in the picturesque villages. Everyone was standing on the balconies toasting and shouting Happy New Year to us.'

Eddie and Pearce got on very well during the trip, but there were times when Eddie's patience was pushed to the limit:

'Gordon handed the driving over to me just before a big mountain climb but he failed to tell me that the two fuel tanks weren't connected. As I got near the summit, the truck just conked out and stopped. It was very cold and snowing.

'Gordon was asleep, so I had to get out on my own and switch the fuel tanks over, then bleed the system through and get the engine going again which took a long time to do. Then the truck wouldn't move because it just kept slipping. I was in sight of the summit and the bloody truck wouldn't move, so I had to get out yet again and put the snow chains on. I was frozen stiff and covered in muck.

'Then Gordon woke up and calmly asked what was wrong. I just stormed off and marched down the mountain. If only he'd have told me about the fuel system, we'd have got over the mountain in one go.'

Brothers Bun (left) and Eddie Parlane pose in the AEC. (REV)

After a couple of runs learning the ropes, Eddie then shared the AEC with brother Bun for two or three trips which kept everything nicely in the family.

On their first trip together the brothers broke down on the German autobahn near Frankfurt. The engine blew up and Eddie said that they had to be towed to the nearest town for repairs:

'We had the complete truck and trailer on tow and the German recovery driver was hitting speeds of over 50 mph along the country roads. I don't think he liked the English. Then we found out that the local garage would not have anything to do with the British-built truck, so a mechanic was flown over from England with all the necessary parts. Luckily the garage owner did allow him to fix our truck on his premises.'

As well as going to Tehran, the Parlane brothers made a trip to Kuwait together. However, the return leg proved to be very stressful as Bun became extremely ill and was unable to move off of the bunk, which meant that Eddie had to do all the driving:

'Bun had eaten some grapes which made him very sick and for my brother to just lie there doing nothing was not like him at all. I had to keep going day and night to get him home and into hospital, so by the time I got us back into Germany I was seeing donkeys all over the road.

'Bun was quite a stubborn man and would not let me stop at any hospital along the way. When we got to the docks in Holland I was so bloody knackered I fell asleep. When I woke up I realised I only had a short amount of time to get the paperwork sorted before the boat set sail so I ran off leaving Bun in the cab. He must have found someone to drive the truck onto the ferry while I had gone and the crew thought it was me, so they started closing all the doors ready to set sail.

'When I came back, they suddenly realised I hadn't got on with the truck so they threw me a rope ladder and I had to climb up the side of the ship which almost killed me. When I got on deck I found Bun standing there watching. He had managed to get out of the cab and asked me why I'd taken so long. That was typical of him. Eventually we made it back home and I got Bun to hospital where he was diagnosed as having a severe case of dysentery.'

During their time with Asian Transport, the Parlane brothers had a good few laughs together and Eddie recalls a favourite story in memory of his brother:

'We had just arrived in Tehran and were unshaven, filthy dirty and very tired. All we wanted to do was get cleaned up and catch up on some sleep, but a British consulate representative met us at the customs compound and insisted that we attend a fancy dress party as he thought we needed cheering up. We couldn't really be bothered but went along to show willing.

'As we walked in, one of the hosts started shouting and clapping for everyone to look at us. She seemed really impressed and commented on what a great effort we'd made.'

The brothers were also lucky enough to be interviewed for a feature about long-distance truck drivers in the June 1968 issue of *Reveille* magazine which declared:

'Two of the drivers are brothers Bun and Eddie Parlane. 37-year-old Bun and 28-year-old Eddie are both married and they laughingly confessed that one of the things they miss most on their long overland treks is the sight of British miniskirts.'

Eddie stayed with Asian Transport for a year before deciding that he really needed to be at home a lot more often with his wife to bring up his own family. After returning from a trip to Iran, he recalls taking the AEC drawbar outfit home. While there, he decided to try his luck for a job at a local transport company and so took the rig with him:

'I drove it straight into the yard and spun it right round on a sixpence, parking it right outside the main office. I walked in bold as brass and asked if they had any jobs. One of the bosses commented on how dirty the rig looked and asked where I'd just come from so I said, calm as you like, "Tehran." The place went dead silent. I ended up working there for the next thirty years.'

Brother Bun stayed at Asian Transport for another two or three years moving on from drawbar outfits to a new rented Volvo tractor unit with which he towed standard 40-foot trailers on regular trips to Kuwait, Saudi Arabia and Doha. Bun's widow Norma has many memories of the stories her late husband related, but she particularly remembers his comments about the people. 'He loved the Arabs,' she said. 'He thought the people were fantastic, it was just the officials that caused the trouble.'

Eddie Parlane takes a last look at the British mini-skirt before departing on another long-haul journey. (REV)

John Frost

Being in the right place at the right time is a phrase which John Frost will never forget. In 1968 he was working for Norfolk haulier Walpole and Wright and occasionally would see an Asian Transport truck parked in the yard. One day and purely by chance, Michael Woodman, who lived nearby, was visiting Frost's employer with a view to buying a trailer which was for sale.

Woodman mentioned that he needed a driver who was 'a cut above the rest' to which John Walpole replied in his heavy Norfolk accent, 'You wotta go und hev a wud wi that bloke acrawst there.' Frost was introduced to Woodman and, after an impromptu chat, was offered a job there and then. 'I jumped at the chance,' said Frost.

Frost's first trip was to Bandar Abbass in southern Iran, sharing a truck with Bob Paul. However, before he set off, he remembers going with Paul to Scotland in the rig, a brand-new Scania Vabis, UHM 25F, to load an electrical transformer onto a low-loader trailer. He then returned to England to load the truck with razor blades which were destined for Tehran. Frost remembers the trip well:

'It was the first time I'd ever driven as a civilian on the Continent although I'd been in the army in Germany for a few years so it wasn't that strange. I'd also done a lot of travelling around the world and that's where I'd got some experience. Bob Paul was a great teacher and looked after me on my first trip. When we got to Tehran, we met a couple of the other drivers at the Hotel Continental where we were all staying which I thought was fantastic. It was more like a holiday than a job.'

Unlike Paul who usually checked a new recruit's credentials, Woodman never asked to see Frost's driving licence when he took him on. If he had, he would have found that Frost actually never passed his HGV test. 'When I left the army at twenty-six years old, I applied for my licence and it came with all the groups on it. I had a licence for everything,' he said.

Peter Cannon

Not long after John Frost started with Asian Transport, Peter Cannon also found employment with Bob Paul's family. Like Frost, he was from Norfolk and had been working on the land and driving trucks for local agricultural haulier Hector Pledger. The work was strenuous and the hours were very long with two or three nights out a week in digs if he was lucky.

One day Cannon spotted a magazine article in Pledger's office detailing a trip to Afghanistan by a doctor and a dentist. This fired his enthusiasm to better himself and get a job driving ultra-long distances which was what he dreamed of doing.

Coincidently Cannon was a friend of Frost's, who had also worked for Pledger, and when Frost came home from his first long-haul trip, Cannon went to visit him and listen to his stories.

Frost said he was going to see Woodman at his house near Attleborough and wondered if Cannon would like to go with him. 'I jumped at the chance,' Cannon said. 'We arrived at Woodman's house in Great Ellingham to find him and Bob Paul playing badminton with their wives in the huge garden. Being a Norfolk country boy I'd never witnessed anything like that before. It was all very pleasant and social.'

A few days later Cannon had a phone call from Woodman asking if he could attend an interview. The two got on very well and Cannon's next step towards his dream job was a road test in one of the trucks. He made his way to Felixstowe where the AEC had just arrived off the ferry. 'I had never driven a truck with a crash gearbox and I made a right cock up of it,' Cannon said. 'I thought I'd blown my chances but amazingly, even though I had no experience, I got the job because I was a careful and considerate driver. I was thrilled.'

Cannon's first trip was with his friend Frost in a second brand-new Scania Vabis, UHM 26F, loaded with furniture bound for Kuwait university. He had a lot to learn: 'I remember remarking to Frosty what a big place "Ausfart" was in Germany. It was signposted for miles. Frosty enlightened me that *Ausfart* was German for "exit".'

Cannon did a few trips with Frost, but was not happy with the pay. 'I was actually earning more in the UK', he said. He made the hard decision to leave and went to work for Walpole and Wright. A few months later Bob Paul contacted Cannon and asked him to come back. This time there was a pay rise and a new truck. 'Needless to say, I jumped at the second chance,' Cannon exclaimed.

Cannon soon became very confident at driving the drawbar outfits and navigating the routes to and from the Middle East, so much so that he put the case to Woodman and Paul that he could do the job almost as quickly on his own and that it would be cheaper for the company to pay a single driver. He also thought he would be able to negotiate another pay rise for himself.

Cannon excelled at his driving duties and finally persuaded his bosses to allow him to go it alone, insisting that it would cost the company less money in wages and it would make hardly any difference to journey times. Much to his delight, Cannon was allowed his wish and spent some good times travelling alone before he was asked by Bob Paul to stop driving and become the company transport manager.

Bobby Vallas

Frenchman Bobby Vallas also joined Bob Paul's family in 1968. Previously he had been working in a circus as a truck driver which is where he met his first wife, Joan, who was English.

Vallas also turned his hand to other duties when the circus was in town. As the trapeze artists descended upside down from the wires, Vallas would catch them at the end of the act. On one occasion he asked a girl if she would marry him to which she replied No, so he dropped her! The next time he asked the same question, she gave the same answer, so he dropped her again. On the third occasion she said Yes so he caught her.

It was while on tour in England that Vallas spotted an article in a magazine detailing a trip undertaken by Gordon Pearce who had driven his truck to Tehran. Vallas was intrigued and made enquiries as to where Asian Transport was located. He eventually found Bob Paul and straight away asked him for a job. An interview took place there and then and a test drive followed. 'It was easy,' Vallas said with his characteristic French accent and chuckle:

My Family

'I'd driven much bigger and longer trucks in the circus. I remember my first trip to Tehran was with Bob Paul and because his mother was from France we got on famously and became very good friends from the first day.'

Vallas was born in Corsica in 1930 but grew up near St Etienne. He has always been a happy-go-lucky man with a gentle nature and a great sense of humour. When I met him, he never stopped smiling, and I imagined that he was liked by everyone he came across on his travels. Although he wasn't the type to offend or upset anyone, a bizarre incident led him to terminate his employment with Bob Paul and move onto European work:

'It wasn't my fault. I was parked in Tehran when all of a sudden I saw two children running towards my truck, then one of them literally bumped into the side of the trailer. I jumped out of the cab to see what I could do for the little girl, but before I could do anything the police arrived and took me to the station.

'I couldn't believe it. They kept me in their jail for ten days. It was terrible in that cell which I had to share with about fifty other men. There were no beds, no toilets and no television and if I wanted anything to eat or drink I had to pay for it.'

Vallas also had to pay the family of the child before he could be released.

'It cost about a thousand pounds and then I was let out. I went straight to the hospital to see the little girl and she was fine. Because I was a foreigner in their country, it was my fault. Their view was that I shouldn't have been there, and if I hadn't been the accident wouldn't have happened.

'The outcome of it all was that the authorities banned me from entering Iran for life which I found incredible considering the incident was so minor.'

Loved by everyone he worked with, Vallas was affectionately known as Napoleon. While he was in Istanbul, he had the name painted on the doors of his Scania, TDX 400K. When he arrived back from his trip, he proudly showed it to Bob Paul:

Bobby Vallas holds his first wife Joan during their time working in the circus in the early 1960s. (BV)

'I fetched Bob from his office and showed him my truck. He pointed straight at the door and said, "What's that?" So in my best French accent I proudly said, "Napoleon".

'"What are you talking about?" he responded. I told him it was French just like me and it made me very proud to see it written on my truck. He just muttered something and walked away so I guess I got away with it.'

Dave Poulton, who worked at Astran in the 1970s, lived near Vallas for a time and clearly remembers him being a big drinker:

'Quite often I'd be at home and there'd be a knock on the door. It would be Bobby and he'd say in his broad French accent, "David, you 'av zee black milk?" We'd end up doing two bottles of Scotch in some coffees in about two or three hours. Then he'd get in his car and go home!'

The Long Haul Pioneers

Ray Scutts

Another new boy to join the family in 1968 was Ray Scutts. Like Pearce, Scutts had come from the army and as an ex-paratrooper was already toughened up for the harsh conditions he would have to face. Nowadays he lives in the warmer climes of southern France and although retired, he has his time cut out looking after his beloved animals and small-holding. 'When you're old and wrinkly, it's the place to be. We have a great way of life here,' he said.

It is clear that Scutts thoroughly enjoyed his time working for Asian Transport and has fond memories of the family with whom he worked:

'I owe a great deal to Gordon Pearce. Without people like him doing the job, a lot of us would have given up on the roadside.

Gordon took me out on my first trip and it was his calm nature and dry sense of humour which got us through. But when you look back now, he should never have lasted as long on the routes as he did because I don't think he had an aggressive bone in his body and you needed to be tough to survive.'

Scutts's inaugural trip with Pearce was in the AEC which he classed as a nice old thing. The two men shared it on a journey to Tehran and back:

'It could really go when it wanted to. If we were lucky we'd get 50 mph out of it downhill. It had a peculiar David Brown splitter gearbox, though. I remember always banging my elbow on the back of the engine hump when I changed gear.

'It was a nice lorry, but it wasn't my favourite. It was very noisy and had no power steering, just a rack and pinion type

Ray Scutts thoroughly enjoyed his days driving for Bob Paul. He is taking a well-earned rest at Bazargan, the border into Iran from Turkey. (RS)

which meant you had to start turning thirty feet before you got there. The only trouble was when you let go of the steering wheel you had to get your hands out the way pretty quick, otherwise the wheel would chop your fingers off as it came back round.'

A stark reminder of the basic, if not crude equipment, that these men had to endure. Scutts also recalled an incident where he and Pearce were chased by a bear:

'Gordon and I had stopped for a piss. We were along the side of the trailer and as I looked over my shoulder I saw this bloody great bear lolloping towards us, so I shouted to Gordon and the pair of us ran to the cab as quick as we could. I never knew I could clear the A frame between the truck and trailer so quickly and I never saw Gordon move so fast either.

'It frightened the life out of us. Gordon leaped up the cab ladder and ended up sitting on the roof while I managed to get back inside. Then when the bear was right next to us Gordon started shouting and banging on the roof to scare it away. After a few minutes it gave up and wandered off.'

Dick Snow

In December 1968 Dick 'Snowy' Snow began working for Bob Paul and made his first trip to Iran at the age of thirty-two. Sadly, Snowy lost his fight with cancer in 1995, so we only have the memories of those he worked with to fill in the details of his lengthy driving career with Asian Transport/Astran.

Ask any of the others how they best remember Snowy and they will undoubtedly recall him always having a large beer in his hand. Snowy liked his drink and made no bones about it. Bob Paul remembers him as always having a gut full of Efes, Turkish beer.

Like Pearce and Scutts, Snowy was military trained but was more used to travelling underwater as a submariner. He joined the navy at the age of fifteen and as it happened was an accomplished swimmer specialising in the butterfly stroke.

After leaving the navy, he chose to drive trucks and, after various jobs, worked for a furniture importer in Hertfordshire which saw him taking his rig to Sweden and Denmark collecting Scandinavian furniture. While on the ferry one day, Snow met up with Bob Paul and the pair began chatting about the Middle East run. Snow got on well with Paul and soon after that chance meeting, started working for Asian Transport.

Scutts recalls taking Snowy out on his first trip. 'We called him Fizzog,' he said. 'It was his hair and big red beard that did it.'

A colourful character to say the least, Snowy never beat about the bush and on more than one occasion got himself and others into more trouble than he bargained for. Scutts recalls one such incident:

'We'd stopped at a restaurant for the night and neither of us had eaten for hours. I wanted a good meal but Snowy was more interested in downing a few beers first so he walked straight up to the bar where two big Swedish drivers were sitting quietly drinking. He pushed in between them and snatched one of their beers, downing it in one. Then he did the other one.

'It didn't look good as the Swedes stood up either side of him. Then Snowy slapped them both on the back and told them it was his birthday and he'd been waiting a very long time for a drink. They all began to laugh and the Swedes actually paid for more beer. The last I saw of them as I went back to the cab, was with more beer in their hands.'

Did they ever get round to having a meal to celebrate Snowy's birthday? 'What birthday?' was his reply.

Life on the ultra-long haul runs was tough and everyone had their own way of doing things, Snowy being no exception. As he gained experience he also gained a reputation for stopping for nothing and soon became known as 'the fastest wheels in the east'. He would consistently complete round trips to the Middle East and back in twenty-two days. The Middle Eastern publication *Aramco World* caught up with Snowy for an interview. The November 1977 edition summed him up perfectly: '17½ stones, six feet two inches tall and looking almost as formidable as his truck.'

Martin Sortwell knew Snowy very well from the days before he worked for Bob Paul. Sortwell never worked for Astran, but used to accompany Snowy on many of his trips to the Middle East and has some very fond memories of his good friend:

'Dick was a real character and I will never forget some of the stunts he used to pull. I will always remember him wrapping toilet roll around his head so it didn't get wet in the snow whenever he went for a crap outside in the winter. It was quite humorous to see it. And if I needed the toilet, he would measure one arm's length of paper and you got no more. He used to ration it off. He would say to me, "If you need any more, you'll have to find a dock leaf."

Snowy decided to strip off in front of a bus-load of German women! The fact that it was well below freezing was of no relevance. (MSO)

'But my favourite of Dick's stunts was when we were queuing at Bazargan on the Turkey/Iran border. He spotted a coach full of German women going back to Munich. It was bloody freezing but nevertheless he said, "I'll give them something to think about," and then stripped right off to his underwear and clogs. Then, bold as brass, he stood in front of the truck and began whooping and shouting and hollering at them until they spotted him and did the same back. He loved the attention. That was typical Dick. God bless him.'

Bobby Vallas was also paired up with Dick Snow on many trips and has very fond memories of the big man:

'We were best mates but it wasn't like that to begin with. I thought he was a big show off when I first met him. We'd stopped for the night at a popular restaurant in Turkey and after a few drinks he began to get on my nerves and became very aggressive. It all ended up with me throwing a punch and knocking him right off his chair in one go. He was a big bloke, but I had very strong arms from the circus work so I took one swing and over he went. He ended up sprawled against the wall.

'When he came round he said, "Okay, Bob. You are not so bad," then I helped him up and we shook hands. We got on famously from then on.'

Drinking was definitely a big part of Vallas and Snowy's daily routine. Vallas said that while they were in Bulgaria once, they were stopped by the secret police who had spotted them in a restaurant:

'We'd had a good meal and a good few drinks. Then we left and a bit further down the road I said to Dick, "We'll have to stop because I can't drive any more. I can't see straight."

'I knew I'd had too much to drink and the next thing I remember was the police banging on the cab door and asking us to get out so they could talk to us. They said we shouldn't have been driving as they'd seen us drinking earlier.

'We got out and Snowy locked the cab, then got upset and threw the key into a field. I couldn't believe it because the spare key was still inside the cab. We all had to get on our knees and search for it. Eventually we found it but by then we were getting on very well with the policemen and it ended up with us challenging them to a game of darts.

'We put our board on the back of our trailer and we all took turns throwing the darts. It was so funny as they kept missing and Snowy was telling them it was they who were pissed, not us. I told him to shut up before we got into more trouble.

'Graham Wainwright and Ray Neal were running with us and they couldn't believe what was happening so they joined in the game too. Then the police took Snowy and me to the station where we had to pay a fine of about twenty pounds for drinking and driving. The policeman told us we should not drive any more and then let us go, simple as that. When we got back to the truck I couldn't resist driving a bit further just to the next lay-by where we called it a day.'

Ruggins, O'Connor and Parkinson

Exact dates are unknown, but in early 1969, three more men began working for Asian Transport: Jeff Ruggins, Brian 'Curly' O'Connor and Brian 'Parky' Parkinson, but all have now sadly died.

Ruggins was a typical East End London boy who had served time in the military, mainly with the SAS, and had previously driven a truck for Turkish Transport & Trading. Word had got back to Bob Paul that Ruggins wanted to work for him, but he thought Paul would not take on a roughneck London boy. Paul was not put off and got a message back to Ruggins asking him to attend an interview. Paul has fond memories of the man and recalls an instance after Ruggins had returned from one of his first trips:

'We were having a drink in the pub when Jeff slapped me on the back which nearly knocked me off my feet. Then he said, "I never thought I'd ever drive a truck for a bloke who spoke proper - and then get to like the fucker." He was very blunt and spoke his mind but it was a nice compliment to me.'

Left to right: Brian 'Curly' O'Connor, Tony Soameson and Bobby Vallas about to try their hand at skiing near Tehran. (PC)

Paul's memory of O'Connor is that he was a lovely, kind-natured man with a heart of gold who originated from Ireland and came to settle and raise his family in Liverpool.

Parkinson died in 2004. His widow, Pam, has some very vivid memories of her husband's time driving to the Middle East and in fact accompanied him on a couple of trips. Before he worked at Asian Transport, Brian was driving trucks around the UK. He heard about a vacancy through his long-time friend Graham Wainwright who was already working at Asian Transport and was able to put in a good word.

Pam accompanied Brian to the job interview where she was quizzed by Bob Paul.

'He assured me that things would be fine and that I could phone him any time if ever I needed to know anything. I thought that was wonderful of him.

'Normally a trip would take about six weeks, but if Brian was ever delayed, Bob would always let me know. The communication was terrible so drivers had to make one call to the office and then they would pass the message on.'

Although Pam wasn't entirely happy with the long-distance trips, Brian agreed to do the job until she'd had enough. Originally, she said, the plan was for a year but he kept at it for two years because he enjoyed it so much:

'The children were growing up fast and Brian was never home to see them. One thing I will say for Bob Paul, though, he always got his boys home for Christmas even if he had to fly them back on Christmas Eve. That was especially nice for the children.

'I remember Brian being picked up from Heathrow airport and getting home in the early hours of Christmas morning. The children were very excited and were all up waiting, so we opened presents and then they went back to bed. He got home by the skin of his teeth.'

Ron Bell

Ron 'Jock' Bell started with Asian Transport in 1969. He had never driven on the continent but he 'just fancied doing something different.' Bell was fed up with trunking from London to Newcastle, South Wales and Scotland in old trucks that would only do 30 mph and he wanted to

Left to right: Ron 'Jock' Bell, Peter Cannon and Ray Neale take an impromptu break for a cook-up. (PC)

My Family

see how they did things abroad. He already knew Curly O'Connor and asked him to put a word in with Bob Paul. Bell was then invited to an interview and took his wife along.

Paul wanted to know what she thought of her husband going away for three or four weeks at a time. 'She told him she didn't mind where I went or what I did,' said Bell. 'So long as I came home to her and her alone, I could stay away all the time.'

Bell's first trip was in a Scania Vabis which he shared with his mate Curly to Tehran. Ironically, Bell's first trip was his mate's last, as Curly didn't really take to the long-distance runs and yearned to be home with his family.

On his second trip, Bell was paired up with Bobby Vallas and after that, took Ray Neale, another new boy, on his first trip.

When trucks were held up for servicing, repairs or were waiting to load, it wasn't unusual to see the boys swapping round and doing trips with different mates or in different trucks. Bell has clear memories of the various trucks he drove:

'Without doubt, the worst truck I drove was not an Asian Transport vehicle, but one that the company was asked to collect in London and deliver to Bandar Abbass in southern Iran. It was a nightmare journey which I shall never forget.

'All the time I worked for Asian Transport, I never had a road map. Curly showed me the ropes and pointed out all the important landmarks along the way. He was a great inspiration to me and that was all I needed. It all sank in very easily.'

Tony Soameson

In the late 1960s Tony 'Fingers' Soameson was working for Lancer Boss forklift trucks as a sales rep but was fed up with his mundane duties. He had seen an article in a *Commercial Motor* magazine documenting the first trip that Paul and Woodman made to Afghanistan. He wanted to be part of their adventure so he wrote a letter

New boy Tony Soameson (right) on his first trip. He shared YYW 330G with Peter Cannon. The pair are seen here resting in Yugoslavia with some al-fresco cooking. (PC/CS)

to Asian Transport detailing his selling experience and asking for a job. He wanted to do some driving to the Middle East for a while and then come off the road and put his sales expertise to good use. He received a letter back from Woodman but unfortunately there were no vacancies at that time.

A few months went past and Soameson had all but given up hope of getting his dream job when another letter dropped through his letterbox, this time good news. He was invited to meet Bob Paul for an interview at Baxter Hoare's warehouse in south London. As Paul was loading one of the trucks there, he thought it would be a good opportunity to take Soameson out for a test drive, but before they could get going there was the small matter of a puncture which needed repairing. Paul takes up the story:

'I noticed that Soameson had two fingers missing and I was curious as to how he would manage if he should have mechanical problems while he was away. I watched as first he put on his overalls, then he got a toolbox from his car and took out the necessary bits and pieces.

'So far so good, I thought. I was quite impressed. In fact he was meticulous and changed the wheel perfectly. I told him that as far as I was concerned that was ten out of ten.'

After the demonstration, a test drive followed which saw Soameson at the wheel of a left-hand-drive truck for the first time in his life. Luckily for him, he passed with flying colours and so began his lengthy career in Bob Paul's family. Soameson's first trip was with Peter Cannon in a new Scania 110 drawbar outfit, YYW 330G, which the pair then shared for about a year.

Knowing that Soameson had sales experience, Woodman asked if he would consider swapping the truck for a sales position. Needless to say, Soameson took him up on his offer and spent the next seven or eight years enjoyably promoting and selling the Astran brand. He was made redundant in 1979.

Soameson's nickname 'Fingers' came from an unfortunate accident when, before Asian transport, he was working for a German machinery company and used to attend their trade shows.

At one particular event in London the exhibits had arrived from the factory but there was no sign of the engineers or demonstrators so Soameson thought he could put everything together in time for the show. Unfortunately things did not go to plan and he lost two of his fingers in the moving machinery.

Geoff Frost

Geoff Frost, no relation to John, started working for Asian Transport in 1970. He had been working for JHW Transport of Camberley and had already made several trips to Tehran, so had good experience of the overland run.

Frost had seen the Astran trucks on many occasions and so decided to visit the Chislehurst office looking for work as he was simply fed up with the truck he was regularly driving and wanted something better. After an informal chat with Bob Paul, he was offered a job which was the start of his twenty-year career with Astran. He remembers that his first trip was to his beloved Tehran. As the years rolled on, he loved hauling trailers to much further destinations including regular trips to Qatar and some to Islamabad in Pakistan and Oman.

Frost said that once the overland routes were modernised in the 1970s and '80s he preferred to drive via Greece to Turkey. This was much more civilised than entering Turkey from Bulgaria at Kapikula:

'The Kapikula route became a depressing hellhole with nothing but queues for miles and miles. Going through Bulgaria and then into Greece via Ipsala was much nicer. That route was a lot longer, but there were no horrendous queues and it was a much quieter border.

'My favourite destination was definitely Muscat. The people were all very friendly and the whole place was so clean and tidy. It was beautiful. I found it extraordinary that if anyone had a dirty car and the police caught them, they would get fined ten riyals and it was the same for dropping litter, ten riyals. What a fantastic idea.'

John Williams

John 'Willy' Williams, also known as 'Diesel Dan', started for Astran in 1971. The previous year he had driven to Rawalpindi, Pakistan, for another company.

Over the next few years Willy became probably the most well known and respected man ever to drive on the Middle East run. He would stop at nothing to offer his assistance to any driver he came across along the routes. It didn't matter one iota to Willy whether they were Astran mates or not. It didn't matter whether it was minus twenty in a blizzard or plus forty in the burning Saudi desert. He was a kind, helpful man who could never see another driver stranded.

Sadly, Willy suffered a heart attack and died in 1996 while he was parked for the night at a truckstop in Germany. Such was his reputation that when his funeral took place in his local village near Northampton, the service had to be broadcast outside because of the huge crowd who had gone to pay their last respects to such a great man. One of the floral tributes was in the design of a TIR plate.

Willy was a good driver, but best of all he was an excellent mechanic. He was posted to Doha when Astran opened a depot there in the late 1970s. First he worked as a driver and then became fleet engineer. Willy was indeed a very likeable character who took great pleasure in telling others about his adventures.

There were some articles in the press which documented parts of his career. The BBC television documentary *Destination Doha* featured the great man, giving a good insight into his lifestyle and thoughts.

Willy was summed up perfectly by the documentary narrator as 'the thinking man's lorry driver', a man who was too experienced to be caught off guard. As Willy put it:

'Knowing that round the next corner is a bad junction or a bad stretch of road or a nasty hill enables you to drive in a more relaxed manner. I was asked by a driver on one of the ferry boats if I carried a map and I said, "No I don't need a map now." When I got off the boat I went into Ipswich and got lost. I probably know the roads out here better than I do in England.'

Never averse to getting his hands dirty, Willy was always covered in diesel and seemed to relish in the substance, hence the nickname 'Diesel Dan':

'If there's something that needs doing, I like to go in and do it. The fact that you're getting dirty is irrelevant. If there's a pool of oil in the way, well there is. Perhaps someone else would tow the vehicle up the road so they weren't lying in oil. To me, I wouldn't bother because oil doesn't mean anything to me.'

John 'Willy' Williams (Diesel Dan) was only happy when he was covered in grease and oil. (DP)

Dave Poulton

Dave Poulton was born to be on the Middle East run. Ever since he was a child all he ever wanted to do was drive a truck long distance and as soon as he could, he was off. He was driving trucks into Europe three times a week before he joined Bob Paul's family and he only got the job because his mate, John Holland, couldn't take it at the time. Holland had waited two years and was poised to start driving for Astran when his wife decided that perhaps she didn't want her husband to be away for such long periods.

Holland mentioned this to Poulton who thought 'why not' and immediately wrote a letter to Bob Paul explaining the situation. He was then invited to meet up with Peter Cannon in London where he had a test drive in a tractor unit. A visit to Paul's house via the local pub then followed where Cannon gave the thumbs-up to Poulton's driving and he was offered a job there and then.

Poulton started in November 1972 on a weekly wage of twenty-eight pounds and his first trip was to Tehran in a Scania 110 tractor unit, ELK 385J. Paul's normal policy was always to send a new recruit out with a more experienced driver but as a load was ready to go and there was no one else available he took a chance in allowing Poulton to go it alone. Poulton clearly remembers Paul's words:

'Bob told me he didn't really like sending new blokes out on their own at that time of year, but if I got to Istanbul and didn't think I could handle it, I was to hang about for a couple of days and Dick Snow, who was behind me, would come along and help me out.

'When I got to Istanbul I thought, "If this is the Middle East I've cracked it." I didn't bother waiting around and carried on to Tehran, tipped, reloaded and came straight home. I found out a few years later that Dick Snow had been going like a train to try and catch me up at Istanbul and show the new boy the ropes - but of course he could never catch me.'

Poulton soon gained a reputation for stopping for nothing: 'Nobody could catch me,' he chuckled. In his first year he only managed six trips to Tehran but altogether completed a staggering 123 trips to various destinations throughout the Middle East in the sixteen years he worked at Astran. He finished in 1988 when he was made redundant.

Dave Poulton completed 123 trips during his sixteen years with Astran. Pictured here in 1975 at the wheel of a brand new Scania, KJN 680P. (ASTC)

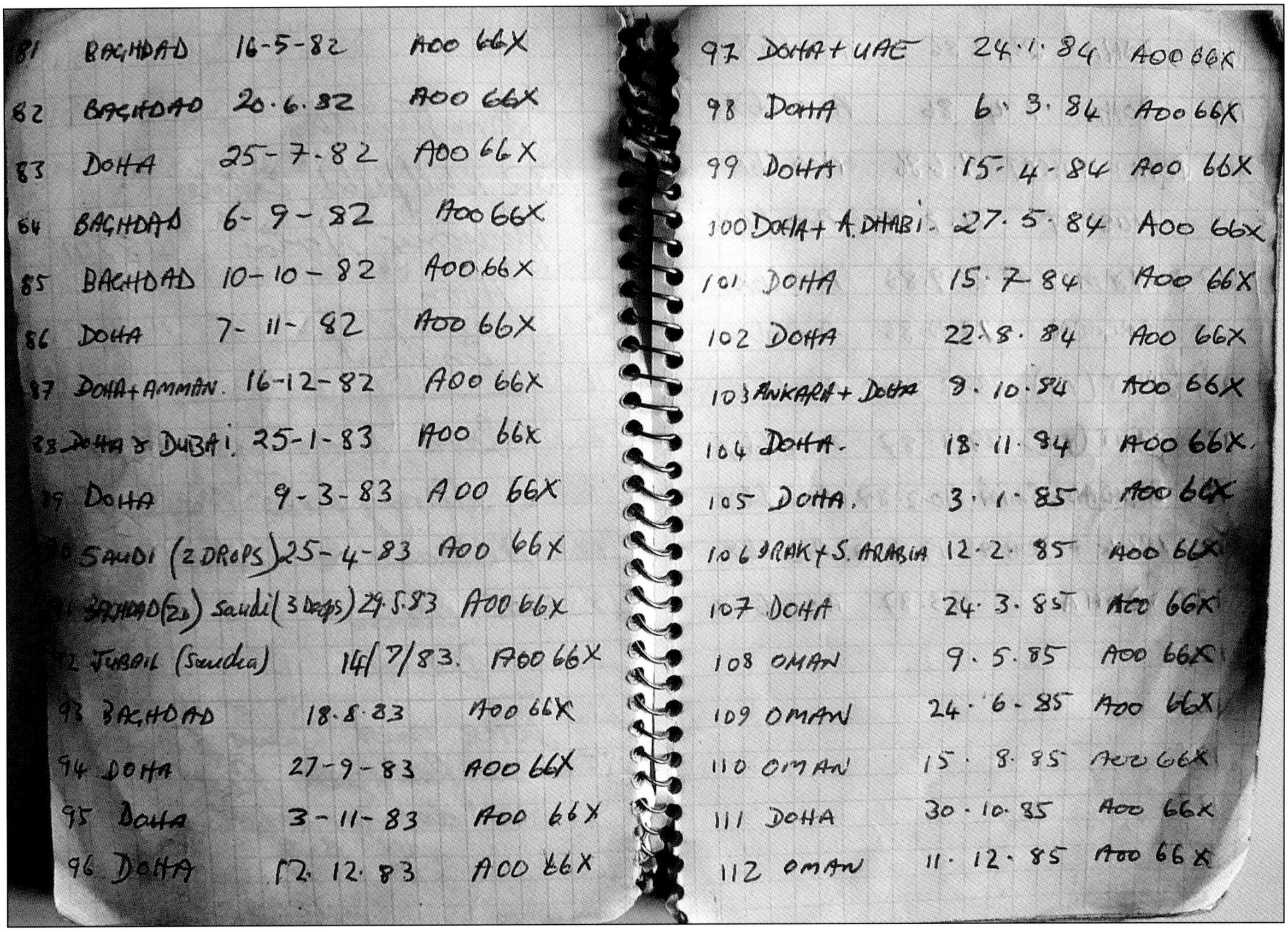

Two pages from Dave Poulton's notebook which lists every one of his 123 trips. (DP)

Poulton has one vivid memory of a chance meeting with boss Bob Paul in Istanbul:

'*I was approaching the Mocamp truck stop when all of a sudden a taxi shot out from the airport junction and carved me up so badly I nearly lost my truck. I put my foot down, caught him up, overtook him, cut him up badly and then pushed him clean off the road. I was furious but by the time I got to the truck stop I'd calmed down.*

'*As I was walking to the restaurant, I heard a very familiar voice call my name and it literally stopped me in my tracks. I looked round to see none other than Bob Paul standing there. He'd flown to Turkey for a meeting with an agent.*

'*The first thing I noticed about him was he was shaking like a leaf, then he calmly said, "Well done, Dave. I could see what was going to happen back there and was ready for a big crash." He was only in the back of the bloody taxi! I thought he was going to fire me there and then but we just laughed about it over a few beers.*'

Charlie Norton

Charlie Norton began his employment with Astran in 1973:

'*I was driving container trucks from Tilbury docks and remember continually having problems with the dockers who were always going on strike. I got really fed up with so many queues and delays and wanted to do something different. I'd driven long distance before the container job so I thought I'd have a go at some European work.*

'*I occasionally used to see an Astran truck in Basildon lorry park, so I took the phone number and gave them a ring. They hadn't got anything at that time but I did manage to start driving for one of their sub-contractors, Morley & Fisher. While I was in Istanbul once, I met Peter*

Charlie Norton (right) always wore his hat. Pictured here with Dick Snow taking a breather in the hot and dusty desert in the mid-1970s. (CN)

Cannon who had flown there to sort out an accident. I gave him a hand with the damaged trailer and then went on my way. When I got home there was a letter asking if I wanted to work for Astran direct. I couldn't say no.'

During his time at Astran, Norton actually left the company on two or three occasions. His wife used to get fed up with the long periods she had to spend on her own so Norton would spend a year or so working in the UK before asking for his job back.

To make things easier for the Astran drivers, they were all allowed to take passengers and Norton was one who took good advantage of the perk. He took his wife and children on one or two trips to Iran:

'Astran was very good to me and if I wanted to take anyone, all I had to do was to ask Bob Paul or Peter Cannon and all the arrangements would be made. It worked very well on the trip that coincided with our wedding anniversary. When we got on board the ferry, the captain called us to see him and told us that my boss had informed him of our special occasion. He then gave my wife and me his cabin for the night.

'We couldn't believe it, and if that wasn't enough, before the ferry docked in the morning, the steward brought us a huge breakfast and champagne which we enjoyed in bed! I have always remembered how good that was. Bob Paul was a terrific boss.'

Terry Tott

Terry 'Totty' Tott was already driving to the Middle East for a Leicestershire haulier when he met Jerry Whelan who at the time was a sub-contractor for Astran. Whelan told him that he had a load ready to go and needed a driver, so Totty jumped ship and began hauling for Astran in 1973-4:

'I'd heard of the Astran reputation and wanted to be part of such a well respected company. I did a few trips with Jerry but I found that he always wanted to stop in Istanbul for a few days to eat, drink and get merry with his mates, whereas I just wanted to get on with the job. I tended to stop a bit further on and mix with different drivers. Unfortunately Jerry didn't like me doing that and when he caught up with me he told me I was sacked.

'I was never one to dump a truck so I continued home and remember seeing Peter Cannon in the office. He asked what

I was going to do next and when I said I wasn't sure, he told me there was a job working directly for Astran if I wanted it. He told me to take two weeks off and then there would be a load ready for me.'

Totty enjoyed his time at Astran immensely and wanted nothing more than to travel further and further. Most of his work was to Doha but he remembers trips to Tehran and Baghdad:

'Bob Paul gave me a trailer loaded for Tehran once, but told me I wouldn't be coming straight back. He said I would be going to Afghanistan to re-load. I thought it was a bit of a novelty, but was chuffed that I was going so far east. I really got into that and the further he sent me, the more I wanted to do it. The furthest I went was to Pakistan with a load destined for India. Unfortunately we were not allowed to go the whole way which was a great pity. I had to trans-ship the load for locals to carry.'

Terry Tott at Ramtha en route to Doha. 'Totty' played alongside – but not in – the pop group Showaddywaddy. (KB)

Apart from driving trucks Totty was an accomplished bass guitarist. Various rumours have been bandied around over the years that he was part of the very successful 1970s glamrock band Showaddywaddy. That is partly true.

It all started after Totty was interviewed by Franklyn Wood who was writing his book *Cola Cowboys*:

'Wood said he had heard that I was a musician so I told him I was a bass player. He then asked who I played with and I misunderstood what he meant so I reeled off some of the groups I had played with. Showaddywaddy was one of them. I was part of the support act but it got put in the book that I was actually part of the group itself.'

Totty and the group all came from Leicester so they were good friends and often seen playing at the same venues and drinking in the same pubs and clubs. The rumour that Totty was part of the band continued and he never actually put the record right himself as it made a good story. Well, Totty, your secret's out now!

Totty's most frightening time was when he was caught up in the Iraq/Iran war during 1980. He said it was that which finally led him to stop driving to the Middle East as he had a young family and he wanted to see them grow up.

Frank Hook

During the early 1970s Frank Hook, working for Kent-based transport company GTI Trucking engaged on long-distance European work, regularly saw Astran trucks on the ferries and at the docks. Talking to the drivers, Hook soon realised that getting on the Middle East run was the next step up from European destinations and really fancied a go at it. Living near Addington, he visited Bob Paul and the pair talked about a job.

In 1975 Hook began working for Astran and during his time there undertook the longest trip that the company ever made. He went to Salalah, an isolated town at the very south of Oman, close to the Yemeni border. The road south of Muscat was about 1,000 km in a straight line and there were only three garages along the whole road. Hook was loaded with road-marking paints. His journey from the UK was supposed to have been recorded in the *Guinness Book of Records* as the longest overland journey ever undertaken at the time, but for some reason the details were never entered.

Joe Bowser

Joe Bowser started working for Astran on 9 November 1976. He remembers clearly that his first trip was to Doha, Qatar, in a Scania 110, GUL 588N. It took him fifteen days to get there and then he spent two weeks working internally in the country before heading home. Luckily, he arrived back in the UK on 23 December, just in time for Christmas.

Bowser had heard of a vacancy at Astran through the grapevine as he was already driving ultra-long distances for another transport company, Haidery Express, which also came from Kent. His trips for Haidery took him regularly to Pakistan so he had experience and knowledge of the routes.

On the off-chance he called into Astran's depot and chatted to Peter Cannon. Luckily for Bowser, he got a phone call a few days later offering him a job. Before setting out on his first trip, he spent some time shunting the trailers and assisting in the workshops so he could get to know the Scania trucks a little better.

It is clear that he enjoyed his time on the Middle East run:

'I loved my days on Astran. There was a lot of trust. We were given a thousand pounds running money and off we went. And if we needed any more, we'd top up on the way.

'I went to Afghanistan and my fondest memory was waking up one morning near Kabul as someone was knocking on my cab door. It was an elderly gentleman from the walled village outside which I had parked for the night. He had brought me fresh warm Nan bread and some tea with fresh goat's milk for my breakfast. What a lovely hospitable gesture. Pity they have such bad press now.'

Bowser stayed with Astran for a year before having to give up his life on the road due to some excruciating back problems:

'I remember my last trip because my back was so bad by the time I got to Belgrade, I had to swap loads with Peter Fanning. He turned round and continued with my trailer and I came back with his load.'

Bowser then worked in the Addington warehouse as foreman until he left the company in 1979. He eventually had an operation to remove a dodgy disc and then at thirty-two years old joined HM coastguards.

Barrie Barnes

Barrie Barnes can boast at being the man who has completed the most trips to the Middle East. In twenty-five years he has made around 270 while being employed directly by Bob Paul, then as an owner driver as a sub-contractor, and more recently with his own business of selling trucks to the Arabs.

Barnes started full-time employment with Astran in 1977 and only has praise for the company which he describes as the best firm anyone could ever work for:

'They never let me down once and they always paid me right on time. They did give me some shit jobs to do, though, ones that perhaps no one else wanted, but that didn't bother me. I'd go anywhere and I did more favours for Bob Paul than I can remember.'

Barnes was trained as a carpenter and joiner. In 1967, before starting to drive trucks, he left England for a contract in Saudi Arabia working for the national air force in the maintenance department. After returning to England two years later, he bought 'various luxury items' with his earnings including a brand new Jaguar car. He met Jean whom he married and then went back to carpentry, working on a new housing estate in his home town of Southampton.

Once he had built his savings up, he decided to buy two trucks to run on a fruit contract for a local company. His fleet grew to three and then Barnes heard of a local freight company which wanted owner drivers to pull trailers to the Middle East. Without further ado he bought another truck which he drove himself, pulling for Abadan Services.

Unfortunately he rolled his truck in East Germany, losing a load of chipboard in the process. After the truck was recovered to the UK and repaired, Barnes contacted Bob Paul and spoke to him about becoming a sub-contractor for Astran. Luckily there was a load ready for delivery to Kuwait which Barnes accepted. Bob Paul also promised him more work, but as Barnes said:

'I was very short of cash at that time and worried that my business was about to fold, so I explained my predicament to Bob and withdrew my offer to take on more loads. Once I got through the tough period I wrote to Bob and explained that I was okay and he contacted me straight away asking me to go back as an employed driver working directly for him. My first trip was to Qatar.

'I wanted for nothing when I started working directly for Astran. I went into the workshop and was allowed anything I wanted for the truck - injector pipes, head gasket sets, valves and so on. There was even a spare half-shaft bolted onto the chassis and, in the winter, the tyres were replaced with new ones especially designed for the harsh conditions. Astran was a proper company run by proper people.'

When the own-account transport operations began to get into financial difficulties during the 1980s Barnes bought himself a truck for £20,000 and continued to pull for Astran for some years as an owner driver.

Barrie Barnes was both an employee and a subbie for Astran. He used this Scania 142, A687 XTC, throughout the 1980s and '90s. (JW)

Barnes is still running to the Middle East but is dealing with 'sellers'. He explained how he got into that business:

'After the first Gulf war in 1990 I was finding times very hard and had virtually no money so decided to sell my truck to an Arab who had offered me good money for it. At the same time he asked me to bring him another one, then another one and so it went on.

'In all the years I've been selling trucks, I reckon I've done about two hundred although I've not delivered them all myself. I always make sure the trucks I sell are in good condition because then I always get more work. On a good trip from the day I walk out of my house, until the day I walk back in I can do it in sixteen days.

'Nowadays there are nothing but motorways and bypasses which speed up the journey times but make the job pretty boring. The early days were definitely the best.

'I remember coming back over Tahir during a bad winter and having to put the snow chains on nine times in one day. I was plastered in mud and dirt and the truck was caked in the stuff. All you could see were the lights, windscreen and number plate.

'When I got back into the depot Bob Paul came out and I'll always remember his words: "Looks like you've had a hard trip." He'd done it and knew what the conditions were. I felt I had achieved something and was back home safe. The load had been delivered and all the bad times were forgotten about, especially when Bob took me round to the pub to celebrate my safe return.'

Kim Belcher

Kim Belcher got the urge to drive on the Middle East run as he walked along a pavement in London:

'I was studying at university and as I walked out of a bank I saw a truck belonging to Oryx Freightlines go past. It had all the bells and whistles fitted and it looked fantastic in the bright orange livery. I instantly thought I wanted to drive one of those. I decided to take my HGV driving test and when I'd passed I found work in the UK for a couple of years.

'Then I was put in touch with a chap from Wales whose company had gone bust in the UK but he was going to ship all his trucks to Doha and set up there instead, so I got a job down there and stayed with him for a year. Then Astran came along and set up just down the road. I met John Williams and got on very well with him and another Astran employee, Edward Doughty. Through them I jumped ship and in 1977 began working for Astran based in Doha.'

Belcher's first taste of the Middle East was actually working internally and it wasn't until a few months later that he moved back to the UK and began driving overland:

'I thoroughly enjoyed that too. It was a great company to work for. It was very interesting and exciting but sadly the job became very monotonous and frustrating at times with hundreds and hundreds of trucks all trying to get to the same place. The queues were horrendous and it was all the hours I spent in those that put me off the job.'

The end of the road came for Belcher when he got so badly drunk in Yugoslavia that he was banned from driving in that country. 'I wasn't interested in driving in the UK after my years on the Middle East run, so I gave up driving trucks completely in 1983,' he concluded.

Kim Belcher started working for Astran at the Doha depot before turning to the overland run. (KB)

Rick Ellis

Rick Ellis just happened to be in the right place at the right time when he started working for Astran. He was doing some work for George Hall, one of the sub-contractors, and was instructed to take the truck back to Astran's depot at Addington and see Peter Cannon. By chance, at the moment when he arrived at Cannon's office one of the regular Astran drivers was discussing his future and asking if he could leave the company without the agreed period of notice. Luckily Ellis was able to start straight away and replace the other driver.

Ellis's first trip for Astran was in 1980. He accompanied Tony Baker and Ray Usher who worked for another sub-contractor, M&C Transport, on a trip to Umm near Basra in Iraq.

Ellis enjoyed his time with Astran and managed an average of nine or ten trips per year until in 1987 Bob Paul informed him that Astran intended to stop operating trucks of their own and he wanted to give Ellis and the other employed drivers the opportunity to become owner drivers:

Rick Ellis at the Jordan/Saudi border with a friendly guard. (RE)

'The last thing that I wanted was to be an owner driver so I considered joining Falcongate, another transport company, but when you've worked for the best it is difficult to work for anyone else. So I thought about it for a while and then took the plunge. I bought a Scania 142, AOO 66X, directly from Astran and continued to work for them for another two years.'

Although he had some very good times working for Astran, Ellis has one particular bad memory which will always remain on his mind. Astran driver Steve Goodman was on his way home from the Middle East and was reloading his rig in Linz, Austria, when he was tragically shot dead. Ellis takes up the story:

'I think it was a Sunday morning when Bob Paul rang to tell me the news. He asked if I could get to the Addington depot as soon as possible and travel out to Linz with Denis McGrath who was scheduled to ship out that night.

'Denis and I double-manned his truck non-stop to Linz and although I cannot recall exactly how long it took us, I do remember entering the city and a taxi driver pulling up alongside us holding the front page of the local paper for us to see the headlines and a picture of Steve's truck. When we arrived at the Chemi Linz factory where Steve was loading I was shown through to the manager's office. I remember him telling me how ashamed he felt that this thing should happen to my colleague in Austria.

'I was then taken to the police station in a Chemi Linz company limousine where I met the officer in charge of the case who explained the circumstances of Steve's death to me.

'Steve had arrived at Chemi Linz at around lunchtime on Saturday and decided to walk to the local bar for a drink and some food. While sitting at the bar he was approached by a "business woman" who asked him to buy her a drink. According to the barman, Steve spoke fairly good German and after buying the drink, he spent some time talking to the woman. Then he thanked her for her company and informed her he must leave at which point she asked Steve to give her some more money.

'Steve told her that she had asked for a drink and she had received a drink. Other than that there was nothing to pay for. Steve did not realise that the woman's pimp was also in the bar and as Steve got up to leave, the pimp called to him and informed him that he must pay, to which Steve replied that he had nothing to pay for. He then turned his back and continued walking out of the bar, at which point the pimp shot him in the back and killed him!

'I did the identification from Steve's passport and was then asked to take a bag of his clothes, albeit not the blood-stained ones, home with me. I returned to Chemi Linz, loaded the trailer and left for England. I was not at all happy being in the truck and I have to admit that I felt like a burglar. While driving I noticed a cassette in the player and pushed it in. Immediately Rod Stewart started singing

Steve Goodman freshens up at a waterhole in Saudi Arabia. (BB)

"My Lady Jane" which was ironic as Jane was the name of Steve's wife. I had only met Steve once before and while I didn't know him well I did know that in life he was highly respected.

'When I got back to Addington, I told Bob Paul that there was a bottle of Chivas whisky and a carton of Dunhill cigarettes in the truck belonging to Steve. Bob replied, "The whisky is no good in the truck. Bring it into the office."

'I was mortified and then Bob said, "Rick, you didn't know Steve so well. If you had, then you would know he would want us to have a drink on him."'

John Bruce

During 1980-1 John Bruce began to work for Bob Paul. Like Joe Bowser, he was already very experienced and had owned his own trucks which were engaged on Middle East work. However, due to a company collapse, Bruce lost everything while he was in Greece and ended up with the miserly sum of just twenty-seven pounds in his bank account. He went to see Bob Paul and explained that his business had folded and that he was in trouble.

Unfortunately Paul had nothing at that time, so Bruce had no choice but to look elsewhere. Luckily he did manage to find employment with the Iranian company Pasagad driving one of their trucks from a depot which they had in Woolwich, London. He stayed there for nearly two years until he received a phone call from Paul asking if he still wanted a job.

During his time with Astran Bruce unfortunately suffered a mild heart attack while in Turkey and had to stay in a local hospital for nine days before he was flown back to the UK. In the meantime another driver was flown out to take over his truck. Bruce had six weeks at home to recover until his doctor was satisfied that he could drive again. He only has praise for Bob Paul: 'He was an incredible boss. That's how good the company was. He looked after me after my heart problems and I will be forever indebted to him.'

I asked Bruce which was his favourite destination. 'That's a difficult question. I went to so many wonderful places, but I guess it would have to be home,' he admitted.

John Lewis

One of the last drivers to work for Astran before the company made all its own drivers redundant was John Lewis. He began his employment some time in 1983-4 after a lengthy career with another Middle East transport specialist Orsan Hak Sertel (OHS) from Rainham, Essex. OHS decided to relocate to Belgium and Lewis jumped ship to Astran after hearing of a vacancy from one of their own drivers. Lewis, who speaks fluent Turkish and some Arabic, was a road foreman with OHS which stood him in good stead with Bob Paul and after an initial interview over a drink and a meal in the local pub Lewis began working for Astran soon afterwards, driving one of the Mercedes Benz trucks.

Lewis clearly remembers his first job which he loaded in Holland with twenty tons of special timber for Oman. The journey was quite slow and laborious due to the density of the timber which

John Bruce. His favourite destination was always home. (JB)

was probably a lot heavier than the documents said. When Lewis arrived at his destination, he found that the timber was to be used in a naval base:

'I was with another Astran driver and we had to stay at the entry border from the United Arab Emirates for about five days waiting for permission to enter Oman. It was so horrendously hot and there was only a makeshift command post with a guard in it and a water tap. All we could do was sit under our trailers begging for the night to come and cool us down a little. As we were guests from England, we had to wait to be invited into the country.'

When Lewis finally arrived at his destination he was pleasantly surprised to be looked after at the British Embassy which had lovely accommodation and proper cold showers. 'That was such a relief,' Lewis said. 'I certainly made good use of them and had about five or six a day.'

John Abraham

John 'Java John' Abraham joined Astran in 1981 as a driver and then moved to the sales department in 1985. The nickname Java John is actually nothing as exotic as it suggests. Abraham was an owner driver before joining Astran and his company trading name was Java Trucking which was painted on his rig. Java John was the obvious nickname, but was actually a mix of his name (John) and his wife's (Val).

Other Drivers

It has been impossible to locate all of the boys. Some of the men I have missed out on are listed below. I am certain there were even more who worked for Asian Transport and latterly Astran at some stage, but sadly I can only offer my sincere apologies to anyone I haven't included in the book.

Verdun Breke
Reginald (Nick) Carter
Edward Dougherty
Eric Ewan
Michael Henley
James (Jimmy) Hill
Paul Hill
Bobby Holland

John Holland (not related)
Anthony Ives
John Jones
John Latham
Denis McGrath
Brian Peak
Roger Pierce
Ronnie Pinkerton

John Abraham. Known to many as Java John. (AM)

Sid Pollendine
Ken Searle
Noel Tehan

Brian Tipper
Mike Walker
Bob White.

The following are those known to be deceased at the time of writing

Royston Henry Day
Steve Goodman
Peter Fanning
Trevor Long
Curly O'Connor

Cyril 'Bun' Parlane
Dick Snow
Ray Thorn
Alan Warner
John Williams

Some of the long haul pioneers in June 2009. Left to right: Gordon Pearce, Peter Cannon, Bobby Vallas, Tony Soameson, John Frost, Bob Paul. (AC)

At a reunion in June 2009 Bob Paul meets his old friend and workmate Bobby Vallas for the first time in over thirty years. (AC)

My Family

Graham Wainwright, the second driver to join Bob Paul's family in 1967. He's still looking fit and well forty years on. (AC)

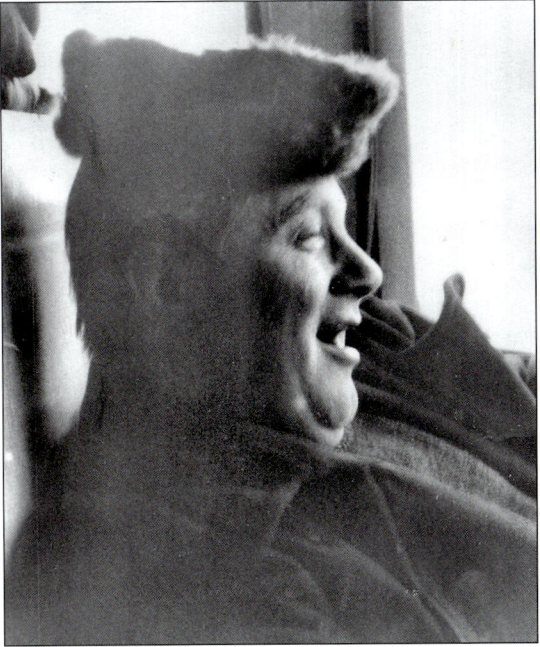

Bun Parlane relaxes in the AEC in about 1968, wearing his unmistakable hat. (BPA)

Bun's hat later found its way on to brother Eddie's head. (AC)

CHAPTER 4

The Extended Family

They came from far and wide, from all corners of the UK, from France, Austria, Switzerland and Denmark. Astran's reputation as an exceptional company - as an employer and as a business - was second to none and anyone who was involved with overland transport to the Middle East wanted a piece of the Astran cake.

To help with Astran's colossal workload many owner drivers and small companies were hired in from the early 1970s. It didn't matter to Bob Paul that they were subbies (sub-contractors), they were all very important to the business. Paul knew them as his extended family, but just like his own employed drivers, he affectionately referred to them as 'my boys'.

It was Paul's responsibility as transport director to set the subbies on and look after them. In an article in *Freight News Weekly* 15 August 1975 Astran founder and owner Michael Woodman indicated the type of driver for which the company was looking:

'One of the much publicised problems on the Middle East run at the moment is the unreliable, inexperienced owner driver. Although we operate our own fleet of vehicles and have a full-

Bobby Vallas (far left) and Gordon Pearce (far right) parked with a couple of Swiss subbies and the wife of one of them. All en route to Tehran. (BV)

72

time team of expert drivers, we are primarily forwarders offering specialised services backed with intimate knowledge of the area and, as forwarders, we regularly contract work to owner drivers. But first, we thoroughly scrutinise them and then, when satisfied, proper contracts are drawn up and on occasion, we finance them fully.

'This is not open to anyone who has a hankering to drive back and forth to the Middle East, but is designed to provide those drivers who really are worth their salt with the means to operate properly. If we are convinced that a particular driver has the necessary driving skills, experience and ability to operate a sound and reliable service, we will buy his vehicle for him and provide any finance he may need along the line.'

Each subbie had his credentials scrutinised thoroughly by both Woodman and Paul. Financial and insurance checks were imperative. Then, so long as the subbie had legal and up-to-date permits and his truck was in good working order with the necessary equipment fitted for the long, arduous journeys, he was taken. This might be for a single trip out, a round trip, or for long-term hire. Most subbies wanted the latter as it gave them secure employment which provided a good regular income.

There were extra payments for those who had trucks painted in Astran's livery, about a hundred pounds per trip for the trailer and another fifty pounds for the tractor unit.

Paul was also very impressed by some Danes and one or two French and Swiss owner drivers who applied to him for work. Marcel Lavanche was one particular Swiss owner driver who stayed with Astran for many years. However, some subbies were not so reliable when it came to the crunch and were soon shown the door. An Austrian driver was given the job of delivering three trailers, one at a time, to the Gulf but abandoned them all en route. He also ran off with the running money which Astran had provided! Luckily, the loads were recovered and delivered intact by more reliable men.

I asked Bob Paul why he used the European hauliers. 'For some reason they could get Saudi Arabian visas very easily and with no problems at all,' he said. 'It made things very amicable between us. The Swiss were the best workers so we used them regularly.'

One of the regular Danish subbies. (BV)

A regular Swiss subbie loading at Astran's Middle East freight terminal, Addington in Kent. (ASTC)

Over the years, there have been numerous owner drivers and small hauliers which have contracted their services to Astran. Some have come and gone again on many occasions, some have stayed loyal for many years and some have decided that perhaps the grass was not quite as green as they first thought and have returned to a more sedate way of life.

There are far too many subbies to mention here and it has not been possible to contact them all, but probably the most famous subbie is Chris Hooper who in 2009 was still providing his services when a suitable load was available for him. Hooper's stories and photographs have appeared many times in transport magazines such as *Trucking* and *Truck and Driver*.

Parked in Qatar, Chris Hooper cleans his beloved Ford Transcontinental while Bob Hedley, another subbie, walks by. (UNK)

John Cooper

John Cooper, probably the very first sub-contractor to work for Bob Paul, started in 1970. He was unloading paper in Enfield one day when he spotted an Asian Transport truck parked nearby so he wandered over and began chatting to driver Tony Soameson:

'I was fed up with general haulage and living from load to load. I was working through the clearing houses and it was very much a cut-throat business with everyone taking their ten or twenty per cent which didn't leave much for me. It was hassle all the time.

'Soameson was very helpful and suggested I ring Bob Paul as he said Asian Transport had lots of work. I did just that and lo and behold I found myself at Chislehurst where I met Bob. We had a chat and it went from there. It was as simple as that.'

Paul arranged for Soameson to accompany new boy Cooper in his truck to Lancaster, where they loaded carpets. Cooper then followed John Frost, one of Asian Transport's own drivers and set out on his inaugural

Hooper's immaculate Scania 143 at rest by the Trans Arabian pipeline, Saudi Arabia. (CH)

trip to Kuwait. Cooper recalls it being fraught with delays at borders, but neither he nor Frost had any breakdowns to speak of which surprised him. Coincidentally, both men were driving Volvo F88s. Cooper's own truck was RBM 510J; Frost was driving one rented from Avis.

John Cooper was the first subbie to work for Astran. In this 1970 photograph he is in Turkey en route to Kuwait. (JC)

Cooper's truck and trailer were already painted blue and white. Bob Paul paid him to have 'Asian Transport' written on it in blue, in keeping with Cooper's own colours, but a departure from the normal red-and-yellow livery. To make his trailer more suited to the ultra-long journeys, Cooper kitted it out with a belly tank and lockers.

Previously Cooper had only ever driven to France and Italy a couple of times, so some proper long-distance work really appealed to him:

'The driving didn't faze me and what I liked best was putting one load on the truck and it staying there for something like three weeks. On the old job, I was sometimes getting three or four loads a day and struggled to get them all off in time.'

Cooper said that he worked for Bob Paul for about two years and thoroughly enjoyed the variety and the unusual destinations. One incident when he and John Frost were en route to Iran still sticks in his mind:

'Bob Paul had sent out a huge box of cut-glass crystal to the agent in Tehran. We found these glasses and thought they were far too good to give to the agent, so we swapped them for cheap ones which we got at the Shell garages when we filled up. When we presented the glasses to the agent we were a little concerned as to his reaction but he was chuffed to bits with them so we said no more. He had probably only ever drunk out of earthenware cups up until then.'

During the early 1970s more and more trucks began running to the Middle East and it soon became a chore queuing at borders so that the trips became longer and longer. Cooper had a young family at home and desperately wanted to be with them more often, so he made the difficult decision to stop working for Asian Transport - although he did continue with some continental work. Cooper told me that Bob Paul was a very good employer who was always fair and paid on time. 'I didn't make a killing financially, but it was a very good experience,' he said.

David Miller

David Miller tried to gain employment directly with Asian Transport. Although he was not lucky enough to get the job, he did drive for one of the subbies during the early 1970s.

He and his mate John were at first turned down by

The Long Haul Pioneers

Bob Paul because of their lack of experience. They got round this by using road maps to mug up on the Middle Eastern routes.

When they re-applied a couple of weeks later Bob Paul was in immediate need for two drivers for his subbies so they were taken on without being asked where they had got their knowledge from. John started with J&T while Miller drove for Eileen Ellingham. Miller said that the work was a dream 'in the amazing early days of overland transport.'

Mike Taylor

Not all of Astran's subbies went overland to the Middle East. Some were quite happy working in the UK tipping and loading the trailers. One such man was Mike Taylor:

'I had a phone call from Peter Cannon who lived in the same village as me. I ran a small haulage company and he had seen my trucks about and asked me if I would be interested in working as a subbie for him. He said Astran was looking for a local operator to move the trailers to and from the docks.

'It fitted in very nicely with our existing work because quite often a trailer would ship in during the evening so we could do a day's work first then go to the docks at night.

'I remember picking up a trailer for another customer and dropping it on a loading bay in Aylesford, then running to Dover, picking up an Astran trailer and taking it back to their yard at Addington, then going back to Aylesford and finding the trailer still not empty.'

Mike Taylor Haulage was employed to tip and reload the Astran overland trailers in the UK. The truck is leaving Dover docks. (DS)

Taylor's relationship with Astran started in the mid 1980s and blossomed over the next five or six years. To begin with he worked from Astran's Middle East freight terminal at Addington. He bought and installed a mobile crane there and was then able to tranship and move the large packing cases and pieces of machinery that Astran were transporting. Taylor has some humorous stories about the depot:

'There was quite a hill going up to the top of the yard where the warehouses were. Between them was a parking area, but it was on a nasty slope and was bit of a pig to get into unless, like us UK boys, you did it often enough.

'We used to take the piss out of the overland boys when we watched them trying to reverse into the awkward bit. I used to say to them they couldn't turn into tight spaces because they were only used to driving in deserts and had a thousand miles of open space to turn round in. They hated that.'

Although Taylor never went abroad he said that he felt like one of the overland drivers at times: 'I remember driving on the M6 when a Frenchman overtook me flashing his lights and hooting at me as he went past. As I looked across at him, I noticed he was grinning and waving a copy of *Cola Cowboys* at me.'

Taylor also had the responsibility of collecting and returning any hired trailers which Astran used and on some occasions found the returning of them to be somewhat embarrassing:

'Saudi Arabian customs were notorious for opening up trailers when they looked for contraband and other illegal substances. On one occasion an officious customs officers had put a forklift truck into the back and used it to tear up the floorboards. In his haste, the driver had got one of the forks hooked under the top rail of the chassis below the floorboards and just ripped it clean off the main chassis beam.

'It was a right mess and the Astran driver ended up with what looked like a giant clock spring at the front of the trailer. He managed to do some basic repairs, but when I took it back I had to explain to the trailer hire company how it had got like that. I really don't think they believed me.

'Another time Astran hired some brand-new refrigerated trailers from Central Trailer Rental (CTR). When I took them back I had to explain why every trailer had huge holes cut into the box-frame chassis. Again, the customs were looking for something and had used oxy-acetylene torches to cut the holes. CTR were not at all happy that their brand-new trailers were so badly damaged.'

Gordon Benn

Cumbrian owner driver Gordon Benn started with Astran just after Christmas 1992 after hearing of the need for subbies through his pal Nick King who was already working there. Benn remembers calling home while he was abroad and being told to ring Astran as soon as possible:

'I spoke to Bob Paul who asked if I could manage a load for him the next day. I told him it was impossible because I was still in Czechoslovakia but was on my way home. I asked him to hold the load for a week and I would do it as soon as I got back.

'Bob impressed on me that it was really urgent, so I rushed back to the UK non-stop, tipped in Harlow then went straight to Astran's the next morning and away I went. My first trip was to Doha and as I'd been there before, I settled into my new job very easily.'

Benn had been pulling trailers for various other companies specialising in Middle East destinations but always wanted to work for Astran. He had made contact with Bob Paul a year or so before, but at that time there was no work for him.

Like everyone else, Benn only speaks highly of Bob Paul:

'What a lovely man he was. He always looked after us subbies and when we went into the offices at Borough Green he would always come out and meet us, shaking our hands and asking how we had got on. If it was lunchtime, he would always take any of us who were there straight to the pub and buy us a meal. It was all very sociable and it's made him so respected.'

Benn had a very good relationship with Astran and explained how he got regular work:

'If you were reliable and quick, they would sort you out with another load straight away. I used to telephone Bob Paul on the way back before I got into Dover and tell him I was on the way to unload at so and so, then I was going to have a week off. That way he would know when I'd be ready for another trip and would normally have something lined up for me. It all worked very well.'

Originating from the north of the England, Benn would normally pick up his loads from there or Scotland. Sometimes he would load out of Rotterdam and even Italy where he recalls loading some very expensive marble which was destined for Kuwait:

The Long Haul Pioneers

'I always remember the particular factory where the office was full of beautiful women and just one bloke who was in charge. He was a lucky bugger.'

Benn was running a 4 x 2 Volvo F12 at the time he started with Astran, but wanted something more substantial, capable of pulling more weight. He was told of a scrap yard in Dorset which dealt in trucks so he paid a visit and found a six-wheeler tipper lorry lying in the corner which he bought at a very reasonable price.

He then made arrangements for Andover Trailers in Berkshire to do the necessary conversion work, chopping off the original back end of his Volvo and replacing it with the tipper parts.

He made a couple of trips with his new truck before deciding to replace the cab and have the whole rig refurbished. Benn was lucky to find another similar Volvo which was almost new but had been involved in an accident and was an insurance write-off.

The Globetrotter cab wasn't too badly damaged and Benn paid £1,500 for it. It then cost another £2,000 to get the truck on the road including having extra large fuel tanks custom-made by another Astran subbie, Roger Haywood, whom he already knew. The whole rig was then sprayed in the distinctive Astran livery.

Sadly for Benn, he was involved in an accident while returning from Dubai which cost him a colossal amount of time:

'I was driving in a Saudi convoy along the tarmac highway when a local truck driver decided he wanted to come past everyone which he was not supposed to do. I was near the front of the convoy when the driver in his big Scania came storming through. Exactly at the point where he was level with me, he had to pull across to avoid hitting an oncoming vehicle.

'It all happened so quickly and even though I was trying to brake I could do nothing to stop it happening. The camel bar on my cab took the brunt of his trailer which slammed into my truck as he pulled in. Nevertheless, the fierce impact tore the cab mounting brackets clean off and left my cab pushed right back and off the chassis.'

Gordon Benn spent approx £3,500 refurbishing his truck to make it more substantial for overland trips to the Gulf States. (GB)

Johnny Neville surveys the damage to Benn's truck. (GB)

'Johnny Neville, working for Orient Freight, came upon me and stopped to help. He kindly dropped his trailer nearby, took mine into the Saudi customs for safe-keeping and then came back for my tractor unit. We put his trailer on the back of my unit and then towed the whole lot back into the customs using his tractor unit so they could see what we were doing and then we headed back to the Qatar border.

'Once we got there, Johnny carried on with his trailer and left me to fend for myself. I hired a local driver with a low loader to take my unit to Doha where I got some locals to try and repair it. They did an amazing job considering the primitive equipment they were using. They had things like toffee hammers which would normally have been used to make decorative bowls and tourist gifts.'

It took Benn about three months to get his truck back on the road. He made a list of everything he needed and then flew back to the UK to source the parts. To save Benn added costs and extortionate import duties, another Astran driver actually hid most of the parts in his load and delivered them directly to the garage in Doha.

It was a risk smuggling the parts into Qatar, but knowing how the customs worked in Doha, it was risk worth taking and it paid off. Getting Benn's truck back on the road and having it re-sprayed, cost him about £1,600 not including the parts.

Benn's original load which had been left at the customs post was picked up by another subbie, Dave Reynolds, and delivered onward to the UK. In turn, Reynolds left his empty trailer with Benn for him to load and bring home when he was ready.

Sadly Benn's trailer was never properly roadworthy or repaired to a high enough standard. Benn had trouble with the ministry of transport which ended up costing him a large amount of money.

To cap it all, Astran's work was dropping off during the mid 1990s. Benn decided to call it a day and stop

The Long Haul Pioneers

his Middle East work: 'It was okay when we were getting eight or nine trips a year, the job paid well, but when work dried up it was hard to keep going. The insurance was costing a fortune too.'

Andrew Wilson Young

This truck driver extraordinaire was a true legend among the men on the Middle East run. Famous for always wearing wellington boots in the winter and plimsolls in the summer, he has now sadly passed away but there are many stories about the great man including a myth that he had royal blood in him. Although that is not true, his father was a Lord Lieutenant of Cumbria which was where Wilson Young got his frightfully posh accent.

Bob Paul has fond memories of Wilson Young who worked for him as a subbie for many years. Sometimes the drivers used to visit Paul and his wife at their house in Kent before they left for a trip, or when they were returning home as it wasn't far from Dover. Paul clearly remembers his wife Caroline tearing Wilson Young off a strip or two on one such occasion:

'The door bell rang bell and there stood Andrew looking typically filthy. We were about to have dinner so Caroline asked if he wanted to join us which he accepted. As he went to sit down she asked if he wanted to have a wash first but he refused. Caroline replied, "Well in this fucking house you do!"

'She never stood any nonsense, even from someone like Andrew, who promptly scuttled off like a naughty boy to the bathroom for a wash and brush-up. I'd never seen him so clean as when he returned.'

Kim Belcher also remembers Wilson Young well and met him on many occasions:

'Andrew never used to sleep. Never! I know that is hard to believe, but I swear it's true. I think he used to drive from

Andrew Wilson Young's first truck, a Mercedes 2023, is leaving from Addington for the Gulf states in about 1971. (ASTC)

Wilson Young very rarely stopped to sleep but he was caught out here taking forty winks over the steering wheel at the Saudi customs. (BPO)

the UK to Doha in about nine days non-stop. He would probably take eighteen days to do a round trip which was unbelievable. He was an amazing bloke.'

Wilson Young never stopped to socialise with any of the others and always preferred to cook up in his cab rather than spend valuable time socialising in bars and restaurants. He was famous for surviving solely on baked beans, porridge oats, rice pudding and yoghurt.

A strong rumour has it that on one particular trip to Kabul, Wilson Young's documents were not in proper order and so he marched into the British Embassy shouting out to anyone that would listen that he was British and he was going to camp out on the lawn until his problems were resolved. This apparently is exactly what happened - and to cap it all, he parked his truck in the ambassador's parking spot!

Wilson Young famously played on his upbringing and on many occasions was able to use it to his advantage. Bob Paul recalls a couple of instances:

'He was waiting to enter Iraq from Turkey and the queue was something like ten miles long just to get to the border post. However, Andrew, being Andrew, just drove straight past everyone and pulled up at the front declaring, "I'm British!" and it was accepted, simple as that.

'His favourite place was the British embassy in Kuwait where he would park his truck outside the front doors and when he was asked to move he would simply reply, "Do you mind, I'm British. I'm staying right here!" He was certainly a character.'

Roger Haywood, another subbie who knew Wilson Young well, summed him up:

'I don't think there was anyone quicker than Andrew. I did one trip with him and never again! I just wanted to see *if I could keep up. Stupid me, I had to take three days off in Doha to recover.*

'As far as I know Andrew was the only one who managed to obtain 'in and out' stamps for transiting Turkey within twenty-four hours. That was some feat in the days when the roads were bad after Istanbul and worse after Ankara. Maybe, if Andrew had slowed down a bit, eaten some real food and had proper rest, he would still be with us now. God bless him.'

Roger Haywood

Roger 'Rita' Haywood began contracting for Astran in 1980. To his workmates, he was known as Rita: '"Rita" was given to me by George Hall, another Astran subbie, because he thought my surname was Hayworth. The nickname stuck from then on.'

Haywood was hauling trailers all over Europe before he chose to try his hand on the Middle East run. To test the water, he did a couple of trips to Greece and Turkey for Davies Turner and freight forwarders Intersped before he joined Bob Paul's extended family where he stayed for about twelve years.

As usual, Bob Paul sent an experienced driver with the new boy on his inaugural trip, so Haywood followed Trevor 'Wing Commander' Long to Damascus and then onward to Doha.

When Haywood joined Astran, the company was being run by the receivers. However, there was still plenty of work:

'We all got paid but had to make sure it was correct. We all knew that Astran still owed lots of money to various companies so we had to be sure that what we got as contractors was the right amount. Personally I never had any problems.'

At that time Haywood began regular runs to Doha which meant transiting through Iraq towards Kuwait:

'I was doing that journey when the Iran/Iraq war was happening and I remember being in Basra when missiles were flying over the top of me. What made it worse was that there were thousands of army deserters from Iraq who were roaming around the area, so we all had to be very careful.

'I remember coming across a driver from another English company, Priors, who was stranded on the desert road. He had been held up by soldiers, one of whom brandished his gun while the others stripped him of everything he had. His cab was gutted and he also lost all his money and clothing,

Roger 'Rita' Haywood converted his truck and acquired a bigger trailer so he could carry more weight and a larger volume of cargo. (RH)

leaving him wearing nothing at all. When I came across him, he was embarrassed but relieved to say the least. We laughed about it afterwards but to be honest the job was quite frightening at times.'

Haywood's truck started life as a 4 x 2 Volvo F12 which he eventually converted himself to a 6 x 2 so he could carry more weight which also made it cheaper to tax.

Dave Poulton

Dave Poulton's early career is described in Chapter Three. After being made redundant in 1988, Poulton continued driving to the Middle East and worked for one or two subbies. These included John Williams, who by then had bought his own trucks, JJ Smith and Trevor Marks who traded as Trailways International.

While working for Marks in the 1990s, Poulton undertook the longest trip he had ever done in all the years he had driven to the Middle East:

'I got a lift to the German/Swiss border with another of Trevor's drivers where I had to meet my truck which had been loaded for me and was left there ready to roll. It was a nice new Volvo FH12 with all the bells and whistles and looked very smart. My truck and the blokes who got me there were loaded for Abu Dhabi.

'After an uneventful trek down there and being unloaded in a reasonable time, I was on the way back to reload in Europe when I had a call on the cab phone asking if I wanted to do a turn-round in Ancona, Italy.

'I agreed to that and met another truck and trailer there at the docks. I swapped trucks and off I went again, this time to Dubai.

'That load was delivered successfully and I was asked to do yet another turn-round to the same place, so back to Ancona I went and swapped again.

'I was on the way down for the third time when Bob Paul rang me. I knew he wanted something because he never rang unless there was an urgent job. "I know you've been out an awful long time," he said, "but could you do another turn-round and go to Doha?"

'Who was I to refuse? So off I went again on yet another turn-round. I was away from the UK for something like fourteen weeks and ended up doing 43,000 kilometres. It was amazing.'

Poulton said that perhaps that wasn't a huge amount of distance considering the number of hours he put in. The

Trailways International was owned and operated by Trevor Marks. Mick Pilgrim is parked by the desert taking some well-earned rest. (TM)

financial reward was well worth it. He earned £7,000 tax free: 'And it didn't bother me in the slightest being away for so long. I loved it.'

Mark Stewart

Mark Stewart worked for Astran from 1992 until 1994 and owned probably the best looking Scania to have ever worn the famous red-and-yellow colours.

Stewart was so intent on becoming a long-distance truck driver that he sold his house and bought his first truck with the proceeds: 'From the age of ten all I wanted to be was a Middle East lorry driver.'

Stewart passed his HGV test first time round at the age of twenty-one and worked for Plackets Transport delivering parcels. He stayed there for eighteen months to gain experience and then took the plunge, selling the house to raise the £12,000 he needed for a Scania 141 tractor unit complete with tilt trailer.

He had found work as a sub-contractor for Whittle International and started with a trip to Athens. His second trip was to Istanbul and then he made regular trips to Ankara. After a time he met up and befriended various English drivers and through one of them, Bob 'Poggi' Poggiani, was able to join Astran. Poggi was

Mark Stewart owned the best looking truck ever to wear the famous red-and-yellow livery. (MS)

already working there and helped Stewart with the necessary permits.

Stewart attended an interview at the Astran offices and after his credentials were checked he was set on. His first trip was to Doha via Iraq just before the first Gulf war started. His Scania performed relatively well but was not really suited to the harsh terrain of the deserts so he part-exchanged it for a better version which had more axles.

As a true truck enthusiast, Stewart had now lived his dream to drive across a desert. His truck was his pride and joy, so he decided to have some modifications and additions done to it:

'First I had some work done at the tank farm in Aksaray where they custom-made a massive camel bar and pair of huge 750 litre fuel tanks.

'I'd seen the shiny American rigs with their big sleeper cabs on TV and wanted something similar, so I went to see a chap in Doha who did that sort of work. I took the whole truck and trailer there and left it for them to work on. It took six weeks to finish and I was so proud when I picked it up. All the time I stayed at a mate's flat and was able to keep an eye on the work.'

The man responsible for the work on Stewart's truck was Kameese Al Khalaf. He was a customs man who dealt with imports to Qatar and so was in a good position to buy trucks and trailers which he then used in his reconstruction business. He also ran his own transport company and employed Indian workers from his Doha premises. It was those same workers who transformed Stewart's truck:

'Khalaf charged me about £2,000 which I thought was a very good deal. I bought all the extra panels, a new roof, new door skins and the other bits and bobs from a Scania dealer in Turkey and then gradually took all the pieces down to Kalefee over a period of a few months.'

Not only was the truck modified extensively on the outside, but the interior was also given the full

Stewart keeps a watchful eye on work at Khalaf's yard, Doha. (MS)

treatment: 'It had a huge double bed and was all done out in red-and-black leather. The icing on the cake was when I had it sign-written. It looked superb with the big red A on the side.'

Stewart's fantastic looking rig also served as a very impressive rolling advert for Astran and became very well known among the other overland drivers as it was so distinctive.

Most of Stewart's trips were to Doha but he did go further along the Gulf peninsula to Sharjah, almost into Muscat:

'I also went south into the forbidden empty quarter. It was a delivery of pipes and it was something like 300-400 kilometres straight out into the desert in the middle of nowhere. Luckily it was a military road and there were checkpoints every so often, so at least I had a route to follow - even though it was just a track.'

Unfortunately for Stewart his beautiful pride and joy was not as reliable as he would have liked. First he had some engine problems and then something far more sinister happened:

'The truck had low oil pressure when I left England, but I just kept checking the oil and the engine. I went down to Sharjah and then reloaded for Moscow. I was sat in the queue to get out of Lithuania and as it was very cold I had the engine running on tick-over.

'I think it was the fact that I was there for three days with the engine constantly running that caused the damage. It was either the big end or the rings, but something just gave up and the engine started knocking like crazy. It was quite serious.

'By chance I met a Swedish driver who wanted to take me back to Sweden so we could get a new engine. He was so helpful which I found extraordinary. His name was Cornelius and we became good friends. I went with him and we found a good second-hand engine at his local Scania dealer. We used Cornelius's car and trailer to bring the new engine all the way back to my truck which I'd towed to a bus depot before I went to Sweden.

Stewart's pride and joy completely burnt out after an electrical fire in Bulgaria. (MS)

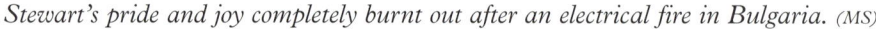

'Cornelius helped me remove the old engine and refit the new one with the help of some equipment and staff from the bus depot. By luck I had picked up a fantastic Swedish-spec engine which seemed so much more powerful than my original one. Once everything was re-fitted and tested, I continued the journey to Moscow and everything seemed fine from then on.'

However, Stewart was clearly saddened by the catastrophe which happened to him a year later:

'The truck caught fire as I was returning to the UK with a load of Bulgarian wine. I was fully loaded with 20,000 bottles and was struggling up a mountain road. I suddenly smelt burning and thought it was the clutch but then realised it smelt more like plastic.

'Before I could do anything, the whole cab was filled with thick black smoke and I had to stop and jump out. It was only by chance that I was next to a lay-by and managed to pull off the road. I just about had time to grab some valuables, but the whole cab was soon well alight and there was nothing else I could do except stand and watch it burn. My fire extinguisher was totally useless against such a ferocious fire.

'I think the cause of the fire was the battery cables which somehow shorted out.'

With help from local passers-by, Stewart was able to contact the police and fire brigade who attended the scene quickly, but sadly there was nothing anyone could do to save the truck or trailer. From then, it was a case of salvaging the load and contacting the insurance company.

Stewart said that it took about three days to get the load transhipped to local vehicles which took the wine back to the winery. The truck was then recovered to a local transport company and left there until the insurance company decided to write it off – and luckily for Stewart they paid him the value of his truck and trailer in full. But this compensation was only financial:

'I was absolutely gutted after spending so much time getting my dream truck exactly how I wanted it. It was a fabulous rig and something I was incredibly proud to own and drive. The disaster really affected me so I decided not to buy another rig after the fire.'

Dave Reynolds

Dave Reynolds was another owner driver who had been engaged on continental work before he joined Astran. He was making regular runs to Greece for Whittle International but the job wasn't ever as good as Reynolds was led to believe. After six months or so he became very disillusioned and moved onto the Middle East side of the company's business with destinations mainly in Iraq.

Reynolds was still unhappy with the company he was working for and decided to make a phone call to Astran. Luckily for him, Bob Paul was in need of another sub-contractor and after an initial meeting Reynolds was offered a trip the following week. Paul was suitably impressed with Reynolds' credentials and the fact that he had his own trailer.

Reynolds thought the Astran set-up was very laid back:

'Everyone was very understanding. If ever I used to phone the office they would always phone my wife and let her know how I was and where I was. It was very much a family business as everyone knew each other on personal terms and they made me feel part of it too. It didn't matter that I was an owner driver, I got treated just the same as their own drivers.'

Reynolds' first trip for Astran was in 1984 when he went to Muscat. He had been expecting to go to Saudi Arabia and had his visas ready, but 'Peter Cannon said I might as well do the load to Muscat. "It's only a bit further," he told me.'

Reynolds stayed at Astran until 1988-9 when he decided to finish driving trucks. He sold his truck and trailer and opened a tool-hire business in Leicestershire:

'The building trade was doing well at the time, but after a couple of years I suddenly got itchy feet and wanted desperately to get back driving to the Middle East. In 1991 I phoned Bob Paul and told him I was fed up with being in the shop all day and asked if I could come back to work for him.

'I was gobsmacked when he said he would have me back tomorrow. That was all I needed to hear, so I went straight out and ordered a brand-new V8 Mercedes and a second-hand trailer which had previously belonged to Andrew Wilson Young.'

Reynolds much preferred to run on his own but also enjoyed meeting up with the others:

'I used to get on with the job. Yes, I enjoyed a good piss-up for a day or two, but then I'd want to crack on. Someone would ask where were we heading for that particular day, but as far as I was concerned I wanted to get as far as possible. If I was running with someone, I was always thinking what they were doing and having to stop unnecessarily for one reason or another.

'I used to meet up with Astran subbie Bob Hedley fairly regularly and I would run with him for a couple of days.

Dave Reynolds was already experienced on the Middle East run before joining Astran. (UNK)

After a spell off the road, Reynolds got itchy feet and bought this new Mercedes 1748 and trailer to begin hauling for Astran again. (SL)

However, he was one of many who didn't like getting out of bed until late. I wanted to do a few more hours so I preferred my own company and I would leave him behind. I must admit, though, he was a good cook and always made a nice dinner and a good breakfast.

'Another time Hedley was following me in Czechoslovakia and I missed a turning. I thought we could carry on, turn right, turn right again and hopefully end up back where we should have been. Hedley copied my every move as he was right up behind me all the way.

'We eventually ended up in a tiny village with a bridge in front of us and a sign saying it was 3.5 metres clearance. I pulled up and Hedley jumped out of his cab calling me a few choice names. I told him he didn't have to follow me and we had a bit of a row about it. We had to do a bit of tricky reversing, but managed to get out in the end, much to the locals' amusement.

'Nevertheless Hedley was a good mate, but he'd never go in front on any of the runs we did together. It took me ages to work out why. If he had gone in front of me I could have blamed him for any problems.'

John Harper

John Harper from Barnsley has been involved with transport to the Middle East for many years, originally contracting to Falcongate, Schenkers and F G Hammond. During the late 1970s he decided to telephone Astran to ask if they had any work. There was a load ready and Harper took it to Baghdad.

The work for Astran continued for a few years until work dropped off and Harper decided to jack it all in to run a taxi business in his home town. In 1988 he decided to start up again and bought a Volvo F10.

John Harper (pictured) put this Volvo to work for Astran but found it was not profitable for him. (JH)

The second-hand Mercedes and DAF were being taken as sellers to the Arabs. The DAF has four trailers – all for sale. (BPO)

Once again Harper managed to get work hauling Astran trailers. The traction work continued for a couple of years until his good friend Barrie Barnes suggested that he start doing sellers.

'I never looked back after that. I was getting into some debt with my own truck and trying to make it pay on the round trips for Astran was almost impossible. When I bought the Volvo I was financially okay but within two years of running it I was £10,000 overdrawn at the bank. My wife told me I was working like a pillock and had only been home seven days in that year.'

Sellers really took off in 1990 at the time of the first Gulf war when Harper was able to make good money taking trucks and trailers to the Gulf to sell on to the Arabs. 'They couldn't get enough of them,' he said.

Some of the other Astran subbies also got into the selling business during the early 1990s. Richie Thorne, Gary Glass, Mike Walker, Barrie Barnes, Mark Stewart, Steve Walters and Bob Poggiani were among the first to do it successfully.

Robert Hackford

Robert Hackford was a schoolteacher for many years, a headmaster for seven of them, who dabbled in transport during the holidays. He gained his HGV class one licence in 1983 after which driving trucks soon got into his blood. He eventually became a full-time international driver.

He was a relatively late starter and his first venture got him half-way to the Middle East, to Turkey in 1995 for Harrier Express of Faversham. After a handful of East European runs he then went on North African work for a number of years, pulling trailers for Breda Transport, Daly Transport Services (DTS), John Mann, Davies Turner and Clive Bull (CSB) into Morocco and Tunisia.

For much of that time Hackford was driving for Jeff Welton Transport of Rochester, which is where he got his first taste of a proper Middle East run. Welton sent him down to the Arabian Gulf on a couple of trips for Astran:

Robert Hackford was delighted with the performance from his Iveco Eurostar, especially the twin-splitter gearbox. (RHA)

'The experienced driver assigned to cover my first rookie trip in 2001 was none other than Dave Poulton. I loaded in Italy and then met him at the docks in Ancona.

'As soon as I saw him, I recognised him from Destination Doha. Even though he had long left Astran, he was still on the Middle East run which I found extraordinary after so many years. We chatted about it during the journey. I thought how good it was to be taken on my very first trip to Doha by this man - and very apt that we were going to Doha itself.

'I saw the route as it had become in 2001, but talking to Dave I was able to get a fantastic overlay of history and how it used to be thirty years before when he made the famous film for the BBC.'

For a brief period Hackford ran his own tractor unit and trailer which he bought from Welton. It was an Iveco Eurostar with an Eaton twin-splitter gearbox which he reckoned to be 'the best thing since sliced bread'.

In the past, Hackford has written many articles for transport magazines. More recently Athena Press has published his Kamyonistan quartet: four novels based loosely on the Middle East run.

Rob Warren

It all started for Rob 'Stutttering Bob' Warren when he was a teenager. He used to stand outside Astran's depot at Addington and take photographs of the different trucks coming and going. 'My favourites were the Turkish Macks. They were fantastic,' he said. Warren was fascinated by trucks and living relatively close made it easy for him to visit Astran.

After a while he became known to the staff and managed to get himself some weekend work washing the trailers. As soon as he was twenty-one Warren took his HGV class one test thanks to subbie Bobby Brown whom he had befriended at the depot and who taught

Warren to drive. (Brown is best known for having been an ITV wrestler many years ago. His tag partner in the ring was Adrian Street):

'I took my test in an old Oryx Freighlines Volvo F89 which I borrowed from a mate of Brown's. The pair were in business together as Double R International. I passed at 10.45am on a Tuesday and by Thursday evening I was on my way to Abu Dhabi.'

Brown's partner wanted some time off, so he trusted Warren with his truck to do a couple of trips for him. Unfortunately for Warren and unusually for Astran he was sent out on his own on his first run. Warren recalls leaving England on that first run some time in early 1975:

'First thing I asked was how I got to the Middle East, to which I was told, "You'll learn, you've got a tongue in your head. See you later, mind how you go." So that was that and off I went.

'When I got to Dover docks, I made my first big mistake. I just followed everyone else and parked the truck in line. Then I ended up queuing for hours on a set of stairs going up to the customs office where I thought the paperwork had to be checked.

'When I eventually gave my papers to the official he asked, "What on earth are you doing here? You have a TIR Carnet so you need to take your vehicle to a different part of the customs building which is round the back."

'I told him I'd just spent two hours queuing and asked why he couldn't stamp my papers. He explained that his office only dealt in general European documents (T forms) and he couldn't help me. I was gutted but learnt a lot that night.'

Warren can always remember his first trip, which maybe was due to the mistakes he made - but nevertheless he enjoyed it immensely. After the ferry docked in Zeebrugge he found his way across to Holland and stopped at the border where he spent quite some time walking around looking for the customs

Rob Warren spent approx £1,500 to buy this Scania 111 from Astran. (DS)

officials because he thought he needed to get relevant paperwork stamped.

Undeterred by all the other trucks roaring past, Warren kept searching for someone until another driver asked what he was doing:

'To my utter amazement, he told me I didn't need to stop there at all and I should simply drive through into Holland. Another lesson learnt.

'I finally found someone going to the Middle East at Aachen on the German border. He was pulling a trailer for Whittle's and was going down to Turkey, so I asked if I could follow him as it was my first time on the run.

'To start with he helped me fill in the German paperwork which was an absolute nightmare to understand. Then he asked me if I had a "Tankshine" to which I replied that it looked rotten and rusty. He laughed at me but I genuinely though he was asking how clean and shiny the fuel tank was on the truck.

'In fact he was asking about the Trakshein, a German legal document which showed how much fuel was in the truck as it entered the country. It was in the cab.'

Across West and East Europe, Warren soon learnt that the best way to drive was very carefully! He was very nervous, especially as he negotiated some of the communist cities whose signposts he simply could not understand.

Eventually, after 'a bloody long time' he made it to Abu Dhabi. Coming back to the UK, he successfully loaded furniture from Romania and delivered it to Northampton without a scratch on his rig.

With diesel now well and truly in his blood, Warren had got the bug and wanted nothing more than to drive permanently on the Middle East run, so after working for Double R International for a couple of years, he took the plunge and bought his first truck, an old Scammell Crusader. He then upgraded to a far more luxurious Volvo F89 and continued part-exchanging his trucks over the next couple of years until he eventually bought an ex-Astran Scania, UAR 747S:

'It was sitting in the corner of the yard looking all tired and dirty and the roof lining was hanging out. I saw Tony Keirle in the garage who started it up for me and I was in love. It had a straight-through exhaust and sounded beautiful. I paid Bob Paul around £1,500 for it.

'Along with the truck I got a huge box folder crammed full of receipts and documents relating to the history of it. It was money well spent and the truck served me very well indeed. Being a Scania, I was also able to make use of Tony's services in Astran's garage.'

Warren mixed his work between Astran, Freighforce of Kent and Fransen of Kidderminster and went to many destinations in southern Europe and the Middle East which he thoroughly enjoyed. One of his most interesting jobs was a load of fish which he took in a fridge trailer from Ipswich to Cairo. The job was done for Fransen Transport, but was sub-contracted to Astran because Fransen didn't have experience of dealing with such a faraway destination.

On the outward journey Warren went overland to Greece and then caught a ferry to Alexandria in Egypt which took about two or three days. Once Warren had delivered to the Cairo fish market, he then had to drive another two days or so further into Egypt to reload potatoes for the UK.

His return journey took him across the Red Sea to Syria and then overland back to the UK. The complete round trip took about thirteen weeks.

Something that Warren managed to do during his time on the Middle East run was to take a driving test which covered him to drive throughout the Middle East. It was not a necessity to have the licence on the Middle East run, but as he was sitting in Baghdad for a long period of time, rather than be bored he thought it would be good to do it:

'It was a farce really and a good laugh. I couldn't understand the instructor, he couldn't understand me and so I just muddled my way around the roads and through a course set up with road cones. He seemed happy with my driving, so after I parted with ten pounds, he issued me with a pass certificate and my photocard driving licence.'

Due to the ever-increasing, extortionate costs and rates levied on trucks in various countries during the 1980s, he began to find it more and more difficult to make any decent money. He decided to stop the Middle East run and concentrate on Greece and Europe instead where the financial outgoings were easier. 'It was a great pity because I thoroughly enjoyed the Middle East work and I loved working for Astran, but I just couldn't afford to run there any more.'

Warren concluded with these lovely anecdotes about himself which some of the other drivers will no doubt remember him by:

'My nickname was 'Stuttering Bob' because of my bad stammer. I used to have a particularly bad time at the Dutch/German border of Aachen because the Germans were

Warren hauled a load to Egypt in this Volvo F89. The job was arranged by Fransen Transport which sub-contracted it to Astran. (RW)

very officious and would scrutinise every single letter and number on the documents and could not put up with my stammer as I tried to explain things to them. The frustration and confusion got worse when I was trying to speak their language. My words just wouldn't come out.

'The drivers who were already parked at Aachen waiting to cross the border knew that I was on a ferry just a few hours behind them, so they would wait for me to catch up. Then they made me do all the talking while they queued right behind me at the customs office.

'The officials used to get really pissed off as they couldn't understand what I was trying to say. The more I stammered, the more frustrated they got, until they just stamped all our documents and waved us through.'

One time Warren's stammer got him into trouble with a fellow-driver:

'As I got to the customs compound at the Turkish border a tyre blew on my trailer. After putting my documents in to be checked, I set about changing the wheel. There were three or four other drivers there, one of whom was a rather large man in a beaten-up old Volvo truck. They all watched me start to work on the wheel when suddenly my wheel brace snapped.

'One of the others told me that the driver in the Volvo had a really heavy-duty brace, so I knocked on his door to ask if I could borrow it. I was hot and bothered and, stuttering like mad, I asked for his wheel brace. Little did I know that he stuttered as well and he thought the others had sent me over to take the piss.

'His reply to me was "Y-y-y-y yes, mate. H-h-h-h here y-y-y-y you are," and he threw it out of his cab window at me. It hit me right on the head and caused a huge gash. To add insult to injury, by the time the others had stopped laughing and wetting themselves, the bloke had got out of his cab and given me a slap. The others tried to explain to him that I really did stutter, but by then he was so angry he took no notice.

'Some good did come out of it though. After treating my own head wound and licking my other wounds, I found the others had changed my puncture for me. When I got back to base, word of my altercation had arrived before me and I was given a piece of slate and chalk and told to write it down next time because it would be safer that way.

'I also remember that if I ever picked up a load for Whitetrux of Dover, I had to send a telex with the details because the transport manager was worse than me at stuttering. He was banned from using the telephone as he used to run up some lengthy and very expensive calls.

'If you stammered, you were thought of as a god by the Turks. They used to look up to you and even today it is something they look on as a special blessing, but it was subbie Bob Hedley who came out with the funniest thing.

'There were a load of us at a bar and he suggested that I should start selling bibles. He reckoned that I would make a fortune. I asked what he was on about: "You could tell them that if they don't buy one, you'll read it to them," he replied. The place was in uproar.'

It is clear that Warren has missed his time on the Middle East run. Even though he has enjoyed the European and domestic work which he has been doing ever since, there has always been something not quite right about it: 'I can tell you that if someone pulled up here now with a proper lorry and said, "There you go, Bob. Fancy a trip to Doha?" I'd be gone within the hour.'

Tony Soameson

As described in Chapter Three, Tony 'Fingers' Soameson began driving for Asian Transport in 1969, but moved on to sales. After redundancy in 1979 he decided to continue with his sales career, and began working for Orhan Hak Sertel (OHS). After an unsettled period there, he moved to Whittle International and finally another move saw him at Falcongate. All three companies were involved in transport and freight forwarding to the Middle East.

Soameson said that none of them were a patch on the operations at Astran:

'I just couldn't drum up the same amount of business because Astran was the name and the brand that everyone involved in export knew and trusted.

'I got so homesick for Astran that I went back to see Bob Paul and told him I wasn't happy and the sales job wasn't working for me. I asked him if I would be able to come back and drive for him as a sub-contractor if I bought one of his trucks. He welcomed me with open arms.'

Soamson bought NVW 480P, one of the Scania 111s. After making some cosmetic changes to the famous yellow-and-red livery he began working as an owner driver trading as Soameson International Freight (SIF) which he had painted on the truck. Unfortunately he didn't realise the significance of the initials at the time and instantly became affectionately known as 'Pox Trans' by all the other drivers!

As work blossomed for Soameson he was able to buy a second truck, this time a brand-new DAF 3300 tractor unit which he drove himself while employing someone else to drive the Scania. After returning from a regular trip in the DAF, Soameson had to make some repairs to the air-conditioning compressor brackets which kept shearing off.

He made the necessary modifications and was busy re-tensioning the rubber drive belts when his assistant, who was working elsewhere on the truck, suddenly started the engine causing the belts to tighten and spin round, wrenching Soameson's fingers into the mechanism! Sadly he lost another finger to add to the loss described in Chapter Three.

Soameson International Freight (SIF) affectionately known as Pox Trans! (TS)

Tony Soameson bought this new DAF but sadly it was written off close to home in Belgium. (TS)

Because he couldn't work for a few weeks, he took on a casual driver for the DAF to deliver a trailer to Muscat:

'I can remember so clearly saying to him before he left, whatever you do, wind the brakes up on the trailer when you get down there because it's a bloody long way and they'll need adjusting. For whatever reason, he never touched them and when he got back to Belgium he couldn't stop and ran straight into the back of another Astran truck, writing the DAF completely off.

'I went to see it at Dover and it was a mess and totally useless to anyone. Luckily for me, the insurance paid out but I had decided to give up the business by then and opened a restaurant instead. That was much more civilised.'

Gavin Smith

Gavin Smith was struck on trucking from an early age. When he was eighteen he used to ride with a mate on trips to Italy but always yearned to get on the Middle East run. As soon as he was old enough at twenty-one he took his HGV licence. Even during his very first job interview Smith explained that eventually he wanted to get on the Middle East run.

European work soon came to Smith and he found himself going to most countries on the continent, but that was not enough:

'When there was nowhere else further to go in Western Europe, I started doing Istanbul and Ankara in Turkey and re-loading in Bugaria and Romania. I stayed on that for several years and finally took the plunge to become an owner driver. My life had been very interesting until then, but now it was to become even more interesting.'

Smith managed to find work with reputable companies but always wanted to haul for Astran. That opportunity came during the 1980s and '90s when he joined Bob Paul's extended family and then spent ten years or so pulling the famous yellow-and-red trailers to the southern Gulf states:

'I loved every minute of it. There were some great lads working for Astran and I made some great friends in my time on the road. I much preferred to run on my own and I've been on trips where, after leaving Ankara, I didn't meet anyone who even spoke English for two weeks or so, let alone see another Englishman or English truck.

'When I finally saw a fellow-countryman, it was always a case of stopping for a brew and a very long chat to catch up on any news. There were also times when I could easily cancel a whole day or two and have a good old drinking session with mates. It was that camaraderie which made the job so good and so special. I met some real good blokes on the Middle East run.'

Smith had two or three rigs in his time as an owner driver. The first was a Scania 141 which he replaced by a better version with Danish specifications around August 1988. He then had the cab rebuilt and the whole rig sprayed in Astran colours.

Unfortunately, in June 1992 Smith had an accident shortly after leaving Ashford in Kent on his way to the docks. He was loaded with cable drums which were quite high and top heavy. As he entered a bend in the road, one of the wheels hit a manhole cover which gave way, tipping the complete rig over onto its side. Smith was very lucky not to have been killed.

He was understandably very shocked by the incident but, unperturbed, he quickly dealt with the formalities and received the full insurance money within a month. He spent some of it on a well-earned holiday and then found a replacement truck, A934 NAR, complete with trailer. He ran this for a year before having it refurbished in Doha, in the meantime doing some sellers while his truck was laid up.

He referred to the new truck as a Scania 'lite' as it had a smaller engine than his previous V8. It was, however, built to Dutch specifications and seemed to pull just as well, if not better, than his old truck. As Smith said, 'It was a fantastic truck. The best thing about it was that I could keep up with the more powerful V8 Scanias even loaded and going uphill. That always made me smile.'

When Smith finished with overland transport there was nothing else left for him to drive for: 'I had achieved my goal and had become a specialist in what I did. I am now living in the Caribbean enjoying every minute – but I still miss those golden years on the dusty road.'

Incidentally, Smith had his initials, GS, imprinted on a ring he always used to wear. Turkey's most successful football team, Galatasary, winners of the UEFA cup in 2000, also wears yellow-and-red colours and just happen to have the initials GS:

'The Turks used to love to see me. I would tell them that I supported Galatasaray and then I would show them my ring. What with my red-and-yellow rig and the ring, it made life so much easier for me at customs and police checkpoints throughout Turkey.'

Proud as punch, Gavin Smith poses with his V8 Scania freshly painted after a refit. (GS)

Smith with his Dutch spec Scania 112 'lite' gives a tow to mate Mark Stewart who was stuck in the sand. (GS)

Roger Gool

Roger 'Rabbit' Gool has very good experience of Middle East work which he started in 1988-9 for Lawarbian. After that company got into financial difficulties, Rabbit bought a truck from them and ran it for about a year until he had an accident, unfortunately hitting a horse which severely damaged the cab.

It was during the rebuild that Rabbit acquired his nickname. His truck was converted to a 'penthouse' rig which included a huge American-style sleeper compartment made from two other cabs which had various panels cut out and then welded together. It was enormous and was affectionately called 'the rabbit hutch' which was painted on the sides. Hence Roger's moniker.

In later years Rabbit had a Volvo FH 12 and a Scania 143 pulling exclusively for Astran for around seven years. When work began to die off in the early 2000s he sold the Scania but kept the Volvo running for Astran.

It was a right-hand-drive truck which eventually meant he could not do any more work with it to the Middle East because after about 2006 only left-hand-drive vehicles were allowed to go there.

Rabbit then found other work with Edwin Shirley Trucking (EST) on rock and pop tours, but always kept in touch with Astran in case a specific job came along.

At the end of 2008, he took a seller to the Gulf which he said he thoroughly enjoyed as he was able to revisit all his old haunts and see some of his old mates on the way down:

'I've been just about everywhere and now I'm on EST I'm going to even more places in more remote areas. It's a real rock-and-roll lifestyle on this job.

'I'd go absolutely anywhere. You've got to, haven't you. If you've been in this kind of job as long as I have it gets in your blood and the further you go, the further you want to go.'

Roger 'Rabbit' Gool. There is no place he would not go. (SL)

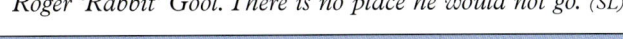

Graham Ball ran this immaculate Scania 143. (RR)

Graham Ball

Graham Ball has been involved with Astran for a long time. He started in 1998 as an owner driver and over the years has built up a small empire. His very successful haulage business, East West Transport, has around sixty-five trucks based in Doha:

'Originally I was working for Davies Turner running to and from Istanbul and I met Barrie Barnes up and down the roads quite regularly. I got talking to him and he put me onto Astran. From then on I've never looked back.'

Ball's main work now is dealing with the sellers. Over the years he has delivered about seven hundred of them.

'To begin with I was sending eight or nine trucks at a time to load for the Gulf, Astran was that busy. The Arabs couldn't get enough vehicles so I was sending all manner of kit down to them. I bought some really ropey old stuff, but it didn't matter, the Arabs bought it all.'

With all the vehicles going to the Gulf, Ball decided to purchase a breaker's yard in Doha which he then turned into a lucrative business dealing with spares for the trucks which the Arabs were buying from him.

In his time, Ball has run some very impressive looking rigs which have been painted in the distinctive Astran livery, the best looking of which was the Scania 143 with high roof conversion.

Nick Bull

During the late 1980s and into the early '90s Nick Bull from Nottinghamshire had a very successful transport business, NB Overland, and for almost a year was engaged on regular Astran work to Istanbul. At the time he was running a second-hand ERF truck which he had converted to his own specification.

Bull was renowned for his talents in the commercial vehicle world and had worked on mates' trucks, including Rabbit's, converting them and adding unusual cosmetic changes:

'Why pay someone thousands to do a job when you can do it yourself. It all moved on from there. I used different parts from many different trucks to get the perfect end result.

'I can't think of a better company to work for than Astran. Money-wise, they were second to none. Once I was unloaded I

Nick Bull Overland. This is one of many Scanias that Bull converted. (NB)

Another Scania 142 run by Bull which served him very well indeed. (NB)

would send the relevant paperwork home and Astran would pay my money straight into the account. It was always there long before I got home. They were always good payers.'

After the ERF and a couple of other less reliable trucks, Bull stuck to Scanias and both trucks shown on the previous page have had major modifications. Both started life as 4 x 2 tractor units but UJW 585Y had the parts from at least three different trucks fitted to it. According to Bull it was the best truck he ever owned and to this day he still wishes he had it.

A820 VVW superseded UJW. Bull extended the chassis and fitted an extra tag axle which gave the truck an extremely long wheelbase. Super-single tyres at the front gave better traction and grip on the rough roads.

Other Subbies

It has not been possible to contact all of Bob Paul's extended family or include photographs of all the trucks. What follows is a brief list of some more subbies and a photographic selection of more rigs.

Alex the German	Albert Henry	RMH Transport
Athol Addison *(D)*	Alan Kendall	Sandtrans
John Brain	Roly Long	Richie Thorne
Alan Bremner	Lumsden's	Mike Walker
Dennis Clemetson	NJ McFarlane	Steve Walters
Eileen Ellingham	Johnny Neville	Jeff Welton
Alfie Foulkes	Pan Express (Billy Hall, Terry Hedger)	Jerry Whelan *(D)*
Gary Glass		Bernard Young
Hanif *(D)*	R&S Transport	
Roger Heath	Pat Riley	*(D) = deceased*

John Williams ran his own trucks after leaving full-time employment with Astran. This Volvo F88 was affectionately known as 'Violet'. (JW)

The Extended Family

Yes, it is Jack Duckworth, but not the one from Coronation Street! (UM)

Bob Poggiani ran this mighty Scammell S26 affectionately known as the Scud. (JW)

Billy Russell, affectionately known as the bionic chicken. (MT)

After leaving Astran in the early 1970s, Ray Scutts set up his own transport business and had this Volvo F88 running for Astran. (RS)

Bob Hedley. (UNK)

Peterlee Trucking. (NH)

Two MANs from M&C Transport parked alongside Peterlee's Volvo F12. (NH)

Gary Lyons. (UNK)

The Extended Family

Brand-new Scania 140 V8 for J&T. (TS)

J&T also ran this heavy-duty Scania 140 V8 6 x 4. (TS)

Caerphilly Van Hire (CVH). (AC)

Bobby Brown traded as Double R International and ran this DAF 2800 as a subbie for Astran. (UNK)

The Extended Family

Ron McNulty ran a couple of MANs in Astran colours. (CJ)

Richie Thorne ran this impressive looking Scania 143. (ASTC)

Trevor Dodwell ran this Scania 113 as an owner driver for a short time. (AC)

Brian Jennings. (BJ)

CHAPTER 5

The Early Years

After Woodman and Paul had successfully completed the contract to Afghanistan they were convinced there was a definite need for freight to be transported by road to faraway places. The three trips to Kabul had proved it conclusively and once word got round that the idea of a driver-accompanied service was not just a flash in the pan, British manufacturers and exporters began to put their trust in the men.

It also became clear to rug manufacturers in the East that this was indeed a safer way of sending the highly valuable cargoes to the West.

Woodman and Paul had struggled against all the odds to prove that the service they wanted to offer was quicker and cheaper than the alternatives. The men came through the experience with a wealth of knowledge and contacts spanning Europe and Asia and they were now suddenly catapulted into a totally new, specialised area of road transport.

They realised that their trading name of Scandsmith was not entirely appropriate, so without further ado they proudly announced themselves to the haulage industry as Asian Transport.

With the promise of work, Woodman and Paul could no longer rely on the little Guy truck, so they made a decision to replace it with something more suited to the vigorous conditions they had encountered. They needed something with better traction, more power, greater carrying capacity and better living space for the crew.

The men visited the AEC factory at Southall in Middlesex and struck a deal to have a vehicle specially built to incorporate Woodman's ideas. They decided that a drawbar outfit – a rigid truck coupled to a trailer – would have more flexibility than a conventional tractor unit and trailer.

Michael Woodman designed the specification of the AEC himself.
The sleeper compartment was fitted out at home by Woodman and Paul. (ASTC)

They chose an AEC Mammoth Major Mk5 8 x 4 chassis because it was strongly made and rugged. With heavy-duty springs and double-drive rear axles it was seen to be the ideal vehicle. The particular truck chosen had an AV691 diesel engine producing 205 bhp.

The truck had come directly from the factory, so it was not exactly how Woodman wanted it. Work was started to convert it to his personal specification. To begin with, the second steering axle was removed and the chassis shortened, in effect converting it to a 6 x 4 truck with a shorter wheelbase of 14 ft 8 in.

Woodman also insisted that the lorry had a sleeper cab. AEC did not build them, so the chassis/cab was sent to Oswald Tillotson at Burnley where they grafted on a 2 ft sleeper compartment. When it was returned to base, Woodman and Paul chose to fit it out themselves with insulation and a bunk. They also fitted a smaller upper bunk, similar to a hammock, which swung down from the cab roof when needed.

Back at the factory, AEC then fabricated a second skin roof above the cab to help reflect the hot sun. They fitted extra-large twin 96-gallon fuel tanks and special desert air cleaners to improve performance.

The vehicle was then delivered to the Norfolk factory of trailer manufacturer Crane Fruehauf to have the truck bodywork and drawbar trailer specially built. A 20 ft container-style body fabricated from ribbed aluminium sheet was fixed directly to the truck chassis. A single-axle 20 ft frameless box van trailer and 'dolly' coupling were also supplied, making 2,500 cu ft of load space available. The total payload of truck and trailer was 24 tons.

Woodman was extremely pleased with this set-up as it was ideally suited for segregating cargoes in the two compartments provided by the truck and trailer. Each container could be separately sealed which meant deliveries to two countries would be possible with minimal disruption to the cargo. Having a dolly instead of a fixed axle at the front of the trailer meant that other types of trailers could be used if the need arose.

Woodman and Paul were desperate for cash to pay for the new vehicle which was costing around £5,000. A consultant doctor at Guy's put them in touch with Robert Gooda, a personal friend and patient. Gooda, who ran a string of squash clubs from Chislehurst in Kent, had recently won a big pot of gold on the Pools. He agreed to invest some of it in Asian Transport.

After his return from the Far East and the first trip to Afghanistan, Woodman was not entirely struck on driving vast distances any more, choosing instead to stay at home and run the business while Paul concentrated on driving. In those days British law stated that a second man should always accompany a drawbar outfit to act as 'trailer brake man', so an employee was needed to accompany Paul in the new AEC. That person was Gordon Pearce who had the honour of becoming the first employed driver of Asian Transport. (See also Chapter 12.)

Pearce's first glimpse of the truck came when he and Paul went to see how it was coming along at the AEC factory. 'I was thrilled to see it,' Pearce said. 'It was a sleeper cab which was unheard of in those days. It had the latest powerful engine with a six-speed gearbox plus crawler. It was double drive and it even had an exhaust brake.'

After a thorough inspection an AEC engineer asked if they had any questions. Pearce remembers:

'In my army days spent in Germany I knew about snowy conditions in winter so I asked the bloke where I could hang the snow chains. He looked me straight in the eye and said we wouldn't need them. He also said the truck had diff-locks and Michelin tyres. He was adamant we wouldn't get stuck.'

Pearce and Paul's next stop was Dereham to see how things were progressing with the trailer and once again Pearce was delighted with what he saw. An expert was on hand to show the men around and also to give a demonstration of the manoeuvrability of the complete drawbar outfit. Pearce was in his element and recalls his conversation with the expert:

'I was very interested in drawbars so I said, "Right, how do you reverse it?" The expert seemed very cagey and said we wouldn't be able to reverse it normally. We'd have to uncouple, turn the truck around and push the trailer backwards using the front bumper.

'I thought that sounded easy – but to connect the trailer dolly to the truck bumper meant getting the jack out and raising the heavy A-frame to the correct height. It all seemed a bit labour intensive.'

Pearce was to find out exactly how difficult it was to reverse the complete rig on his very first trip to the Middle East. In later years, Pearce would become well known for his ability to reverse a drawbar outfit onto a sixpence. In fact he became expert at it.

The Early Years

Gordon Pearce was given basic instructions on how to reverse the rig at the Crane Fruehauf factory in Norfolk. (GP)

With the much-needed extra cash from Gooda and the new truck about to go on the road, Asian Transport began to expand its horizons with confidence. In London another deal was struck, this time with Conemar Limited, an international shipping and freight forwarding company. There was a strong business connection between Britain and Iran at this time, the mid-1960s, and Conemar and Asian Transport co-operated to set up a scheduled overland road service direct from London to Tehran.

The idea was to beat the rail/road service which was taking around thirty days, including some time-consuming transhipping and re-loading en route. Having sealed TIR containers and a driver-accompanied service throughout the journey made the consignments much safer.

Other benefits included full warehousing and customs facilities at both ends. Through Conemar it was agreed that Asian Transport would use the Iran-Turkey Transit Company in Tehran as clearing agents. At the time of the inaugural trip the press reported Conemar as saying that this was 'An ambitious and adventurous gamble based on the conviction that this service can be of immense value to exporters in their marketing campaigns.'

Woodman and Paul had negotiated a price of £600 for Conemar to use the drawbar outfit as they saw fit. A date was set for departure to Iran but, as it approached, it became clear that there would not be enough cargo to make the trip worthwhile so another week was allowed to drum up more business. Unfortunately nothing transpired so the drawbar outfit began its journey only partially loaded with just six tons of cargo. Best described as exciting, harrowing and frightening, the trip can be read about in detail in Chapter 11.

During his trips to Afghanistan, Paul made some important contacts which he hoped he could rely on for future work as the Conemar job didn't pay as well as was originally expected:

'I was introduced to Rhatan Singh when we were passing through Tehran. He was the import shipping manager at the agent for Johnnie Walker whisky. I spent goodness knows how many days convincing him that he could load the whisky on our truck in England and we'd take it directly to him without any transhipping, damage and with minimal pilfering.'

The normal route for freight in those days was in sea

A lucrative contract was hauling Johnnie Walker whisky to Iran. Here it is being offloaded at the customs compound in Tehran. (PC)

containers from England to the southern Iranian port of Abadan Khorramshahr and then in local trucks to Tehran. Using this method, there was always a great deal of damage and pilferage mainly at the docks which was costing the Iranian importer a huge amount of money. Paul continued: 'I eventually managed to talk Singh into using our service. I was delighted and I think we got one shilling per kilo which was good money then.'

The whisky contract eventually became one of Asian Transport's best jobs. Once the work was in full flow, pardon the pun, the rigs were carrying around forty tons to Tehran every month. They would load up in the customs bonded warehouse at Felixstowe, get the truck sealed there and then and drive straight onto the ferry.

Sometimes it was more difficult than others to load up, especially if the trucks were already part-loaded with other cargo. When they arrived at the warehouse to top up with whisky, the dock workers would take one look at the small amount of space left inside the trucks and run a mile. Paul explained to me that on more than one occasion he would find himself crawling along the tops of crates and boxes to try and squeeze an extra layer of whisky into the trucks. 'Remember,' Paul said,

'we were getting a shilling a kilo. We had to use every inch of space.'

Another lucrative job in the early days was for Gillette, shipping their razor blades to Tehran. They were branded as 'Nacet' in Iran and the quality seemed to disappear as soon as the name changed. 'They were fantastic,' Paul explained. 'Half of your face would be shaved nicely but they couldn't cope with the other half and it would be left very rough, they were so cheap.'

Ray Scutts also has fond memories of the blades: 'They were stainless steel and if you left them outside for just a minute, they'd go rusty.' One particular occasion he remembers well:

'I was sharing a truck with Dick Snow and we got to Kapikule on the Bulgarian/Turkish border during Ramadan. The officials were in a rather unpleasant mood and wanted to know exactly how many blades were in each box. Neither of us knew, so they decided to open up and count out a few boxes.

'It was obvious that we would be there for quite a while which I wasn't impressed with, but Dick was delighted. It was a convenient place for him to have a drink so he

wandered back to the Bulgarian side of the border and bought a whole load of cheap local hooch. At least one of us enjoyed ourselves while we waited.'

It wasn't at all unusual for trucks to be pulled up at borders and indeed for the whole consignment to be unloaded and counted by the bureaucratic officials. Gordon Pearce remembers this well and relates another story:

'We used to load in London and pack the boxes of razor blades so tightly to maximise the load space, it was a real hard job to fit them all in. After the Iranian customs men had spent hours unloading and checking them all, I wondered how they would fit them back again. They didn't make a very good job at all as they pushed and squashed the boxes in. There were gaps all over the place, but somehow they all seemed to fit which I thought was a bit odd.

'Then one day the customs officials tried to stop me entering the office, but I did manage to get a quick peek inside and there behind the door were at least a dozen boxes of unopened razor blades. That solved the mystery of why the customs men were always clean shaven.'

Without doubt the most important customer for Woodman and Paul in the early years was Oriental Carpet Manufacturers (OCM). Owner Brian Huffner ran the company from his premises near St Paul's in London. He had agents all over Iran and Afghanistan who sourced the very expensive hand-made carpets and rugs from their respective areas.

Huffner soon recognised that trucks could bring cargoes back to the UK far more quickly and safely than any train or ships so it wasn't long before he agreed for his precious carpets to be loaded only into Asian Transport trucks.

With only one truck during the first half of 1967, Asian Transport coped extremely well as the AEC ran exclusively between London and Tehran. Bob Paul shared the rig with Gordon Pearce while Michael Woodman tried to drum up more business in the UK.

Spring 1967: Bob Paul (foreground) is supervising one of the very first return loads of Persian carpets at OCM's warehouse near St Paul's in London. (CAR)

With ever-increasing loads coming from OCM and other customers, it was obvious that another more powerful truck was needed to relieve the poor old AEC of some of its duties.

Woodman wrote a letter to all the truck manufacturers telling them where he went, outlining his requirements and asking what they had to offer. He had always had a plan to use only British-built vehicles, but he did not receive a single reply from any British truck manufacturer. It seemed there were no trucks available that would stand up to the harsh conditions of his journeys.

However, Woodman did receive a reply from Swedish manufacturer Scania Vabis: 'The way our vehicles come off the factory line is exactly how you want them. They will go anywhere in the world. We guarantee it.'

On reading this, Woodman immediately ordered two brand-new trucks for about £16,000 each. That decision heralded the start of a fruitful relationship with Scania trucks which lasted until 1988 when Astran closed down the own-account transport department.

To cope with the mixed variety of loads, they decided to continue to use the drawbar outfit proven on the AEC. The new trucks which Woodman ordered were fitted with 'twist locks', a simple but very effective way of locking containers onto the truck/trailer. This meant that standard container bodies could be easily swapped around, or left off completely if an oversized load was to be carried.

Bob Paul clearly remembers his first visit to the Scania Vabis factory at Sodertalje in Sweden to see the trucks being built:

'I tried to explain to them what we needed and told them they didn't know the conditions we endured, but I got a very frosty response. They were very arrogant and said, "You don't tell us. We tell you what you want."'

There was also a special testing area at the factory and Paul found himself invited to see it which was a remarkable occurrence as it was, and still is, top secret. Again he recalls the experience:

'In fact they did have it set up to the conditions we were used to in Turkey. I said to them, "Yes, you do have it quite right. The only snag is yours lasts for a quarter of a mile and we go the best part of one thousand and five hundred miles." They weren't sure what to say after that.'

As the delivery date for the new vehicles approached, it became clear that they wouldn't be ready in time. Asian Transport already had work scheduled for them so, after much discussion, the local Scania Vabis dealer agreed to lend a similar vehicle until the new ones were available. Although SVB 300F was a brand-new demonstrator, it had no sleeper cab. This made life extremely difficult for Paul and Pearce when they took it on a round trip to Iran to assess its qualities while the AEC was being driven by two other employees.

Even though the demo truck had no sleeper compartment, the cab had much better springs than the AEC which made for a far smoother ride. As one man drove, the other was reasonably comfortable in the passenger seat. Sleeping was very awkward, but with a good cab heater, a couple of blankets and a pillow, it was a case of catching forty winks as and when.

The best time for sleeping came at border crossings where there was plenty of time spent waiting for documents to be processed and loads to be checked. A quick service and some running repairs to the rig could also be carried out during the wait.

Mailshot to potential customers outlining Asian Transport's services. (ASTC)

Essential items such as food and clothing were kept to a bare minimum and crammed into every nook and cranny in the cab. Tools and spares were stowed in a small locker mounted on the trailer chassis. Pearce compared the Scania Vabis demonstrator to a Rolls Royce after driving an Austin Seven, that being his AEC.

Best of all, the truck had more power and speed than the AEC which enabled the men to improve greatly on their journey times. 'Even though I had a great fondness and loyalty for the AEC, I suddenly fell in love with the Scania Vabis,' Pearce said.

Pearce's favourite feature of the truck was the lovely air-operated handbrake mounted on the dashboard which was operated easily by a quick flick of the wrist. The AEC handbrake, on the other hand, was a heavy

Scania Vabis, SVB 300F, and Leyland Beaver, BLH 764B, loaned to Asian Transport while new vehicles were being built. (GP)

Bob Paul (right) and Gordon Pearce spend time tidying the very cramped cab of the Scania Vabis. Note the makeshift foam mattress tucked behind the cab. (GW)

It was a dirty job, but someone had to do it! Bob Paul (left) and Graham Wainwright make repairs to the Scania Vabis. (GW)

Servicing and running repairs were usually completed while waiting at a border crossing such as this one at Kapikule on the Bulgarian/Turkish border. Note the portable washing line and makeshift curtains. (GP)

The 'pink ladies', UHM 26F (left) and UHM 25F were streets ahead of the AEC - or indeed anything else British-built. (JF)

ratchet handle linked to a long lever which needed at least four or five hard pulls, with both hands sometimes, before it would hold the truck in a stationary position.

Alongside the Scania Vabis demonstrator, the dealer also agreed to lend a three-year-old Leyland Beaver tractor unit, BLH 764B, and trailer to help fulfil Asian Transport's commitments. It had been part-exchanged by a customer for a new truck and had already seen a good life working throughout the UK on general haulage.

Graham Wainwright was given the honour of driving the Leyland to Tehran on his first trip in 1967, his trailer loaded with a large razor blade machine. He followed Paul and Pearce who were in the demonstrator, escorting him on another of their regular runs to Iran.

1968 saw the Asian Transport fleet finally grow with the arrival of the two new Scania Vabis rigs, badged as LB76, but to cope with the ever-increasing workload, it was decided to keep the AEC in service and run it alongside them. Everyone was suitably impressed with the new trucks, UHM 25F and UHM 26F, and they soon became affectionately known as 'the pink ladies' due to their factory-painted pinkish-red cabs.

Like the demonstrator, they were definitely streets ahead of the AEC - or indeed any other British-made truck. The main difference was the turbocharged engines which allowed faster journey times, especially in the high, mountainous regions of Eastern Turkey. The AEC with its non-turbo engine would simply 'die' as it struggled for air in the high altitudes.

The Vabis was far more luxurious, as well. 'For a start you could hear yourself talk and when you closed the windows it was lovely, no draughts,' Paul said. 'And they had proper heaters too!'

With properly sprung and adjustable seats, the cabs also had two comfortable bunks. In fact the complete truck was ideal for the ultra-long distances that the men were going to endure.

Paul was very impressed with the new vehicles. On his earlier trips he had asked many Turkish drivers which trucks they preferred. Unanimously they had always told him, 'Vabis is best'.

*'Vabis is best'. UHM 26F posing near Basra, Iraq, en route to Kuwait.
The grille has been removed for better air flow to the engine in the desert heat. (JF)*

'I think that was what influenced Mike Woodman and me the most,' Paul said. 'They were the finest truck out, even if they were the most expensive. If the Turks got on with them that was good enough for me.'

Peter Cannon drove UHM 26F for a time and remembers the attention he used to get from the Turks:

'They would always look at the Vabis with awe and say with gravity "Vabis." They were very impressed with it as the only ones they ever saw were the long-nosed bonneted versions. They were few and far between and very expensive for them to buy, so it was always good for a pose whenever we could. This stopped happening so much when the manufacturer's name was re-branded to just Scania, losing the Vabis. The Turks would still ask if it was a Vabis.'

Even with regular money coming in, there was still a need for more financial help. At this time Woodman was living in Attleborough in Norfolk and he used this as the first registered address for Asian Transport. He also owned a dental practice in Norwich which he shared with Scotsman Tom Gardiner. Woodman convinced his colleague to put some money into Asian Transport and become a silent partner.

Paul's wife Caroline didn't have any part in the company. 'However,' Paul said, 'she gave me much needed emotional support and did draw out her pension from Guy's hospital where she had worked. I think it was about three or four hundred pounds which was enough for a trip back then.'

Woodman continued his dentistry alongside running the company, and when possible would relieve a driver at the docks, taking the truck to be unloaded, serviced or readied for the next trip.

On one occasion Woodman took one of the new Vabis trucks to Scotland to load the trailer with an electrical transformer bound for Iran. On the way back down in England he loaded the container. He telephoned Paul and explained that it was 'a bit of a heavy load, but not too bad to drive.'

When John Frost, who took it to Iran, pulled into the docks and weighed the rig, he was somewhat surprised to see that the total weight of the truck, trailer and load was almost 57 tons. In those days the legal weight limit for a drawbar outfit in the UK was 32 tons!

On another relief job, Woodman broke down and was

UHM 26F resting in Eastern Turkey en route to Iran. Fully loaded, the rig weighed about 57 tons! (JF)

befriended by Jim Tugby who pulled up to see if he could help. Coincidently, Tugby was a diesel mechanic with premises near Attleborough so Woodman seized the opportunity and arranged for Tugby to service, repair and look after the trucks between trips.

Before he met Tugby, Woodman had relied on Essex Scania Vabis dealer Scantruck and AEC in Middlesex to look after the vehicles. Both companies were very expensive but at the time there was no one else who Woodman could rely on. Tugby's work was so good that Woodman decided to use him instead of Scania and AEC, an arrangement that continued for many years.

The round trip to Iran and back was almost 8,000 miles, so it was necessary to service and repair the trucks along the way. Luckily, Paul had previously met Englishman Harry Sully who was living and working in Tehran. A Lancastrian, Sully was in charge of British Leyland's Tehran bus depot.

On his inaugural trips Paul had met Sully and became very good friends with him and was able to use the garage facilities. That meant that he could have the trucks and trailers serviced and repaired at a very reasonable price before they set off back to England. Sully benefited too, because he was able to let his local apprentices get to grips with different sorts of vehicles.

The arrangement lasted for many years until Sully decided to return to England. By then there were more truck dealerships setting up around Europe and the Middle East and it was not so difficult to get repairs and servicing done. Paul was able to set up accounts at various Scania garages in Holland, Austria and Turkey.

Later in 1968 the trusty AEC was finally put to rest and part-exchanged for the first brand-new Scania 1 series which was the successor to the Vabis LB76. The new rig, YYW 330G, had a much squarer and bigger cab than anything before. It also had a more powerful 11 litre engine with 275 bhp and it had ten gears.

All these characteristics appealed greatly to everyone - especially the roomy cab which was fitted with properly suspended seats and two big comfortable bunks. YYW was delivered ready painted in the latest Scania factory-applied livery of a red chassis and yellow cab with a red band and roof.

Woodman was so impressed with the striking colours

Harry Sully, an excellent mechanic, was in charge of the British Leyland garage in Tehran. Sully, holding his daughter, is with Bun Parlane. (BPA)

that he decided to adopt the livery and use it on all his trucks from then on. To this day Astran still uses the distinctive yellow and red on its trucks and trailers, albeit in a more modern version. Just like its predecessors, YYW was also a drawbar outfit which had been built to Woodman's tried and trusted designs.

On YYW's first trip Peter Cannon was given the honour of sharing it with Tony Soameson. Cannon remembers running it for two or three more trips before it was re-registered to AMY 147H:

'We arrived off the ferry at Felixstowe one day to find Mike Woodman waiting for us. He presented us with a set of new number plates bearing the different registration mark and told us to swap them over there and then. We never queried anything, just did as he asked, but I still wonder to this day why it was changed.'

Soon after YYW arrived, a second new Scania 1 series was delivered. Also a drawbar outfit, it was registered as WLO 95G which was never changed.

The brand-new YYW 330G on trade plates. After a couple of months in service the truck was re-registered AMY 147H. (ASTC)

Ray Scutts was charged with WLO and, like Cannon, was promoted from a Scania Vabis. Not only did Scutts have a new truck to drive, he also took new boy Dick Snow on his first trip to Tehran during the winter of 1968. That journey was not without its problems and Scutts recalls a particular incident on their way home through Turkey when the new trailer nearly burnt out:

The Long Haul Pioneers

New Scania 1 series WLO 95G with Verdun Breke driving alongside an older Vabis, UHM 25F. While they take a rest in Turkey Dick Snow is watering the flowers. (RS)

'The whole rig was sixty-odd feet long and it was very difficult to tell if there was a blown tyre on the back axle of the trailer, especially as we were driving at night. I sensed that something wasn't right as the trailer felt like it was swaying too much. We stopped to investigate and found that one of the rear wheels was actually on fire.

'Dick and I threw handfuls of sand to extinguish the fire but the wheel was so hot it just kept burning. A bus stopped and all the passengers got out to try and help. To be honest none of them had a clue and all they seemed to do was run around shouting and waving their arms frantically. I thought that if we didn't do something quickly the trailer would catch fire and then no one would have been able to do anything at all.'

As it was, Scutts's rig was loaded with very expensive Persian carpets for OCM and he knew he had to do his best to save the cargo:

'I had no choice but to break the customs seal off the doors and get as many bales of carpets out as possible before it was too late. If I couldn't save the trailer, at least the valuable load would be safe. With Dick's help we worked like fury and actually managed to unload all the bales. Then as we unloaded the last one, we watched the fire burn itself out.

'It was getting light by now, so all we could do was fit a spare wheel and tyre and then reload all the bales. They were very heavy and we were totally exhausted by the time we'd done all that.'

When the pair arrived at the Turkish exit border to Bulgaria, they then faced a big problem with the authorities. Scutts had broken the sacred customs seal to save the load, something that was not allowed under any circumstances. The customs officer was not impressed.

Scutts was at his wits' end and very frustrated about the problem. He was tired, filthy and somewhat singed as he stood in front of the customs officer explaining exactly what had happened and what he and Snow had done to save the cargo. The official was not in the slightest bit interested and told Scutts he should have contacted the police so they could have witnessed him breaking the seal. Scutts was exasperated:

'I ask you! It was 3 am, pitch black, we were in the middle of nowhere and the trailer was about to go up in flames taking half a million pounds worth of carpets with it. Where on earth did he think I was going to get a bloody policeman from?'

While driving through the night, Ray Scutts and Dick Snow nearly had their trailer burnt out. (RS)

It became obvious that Scutts was getting nowhere with the authorities, so he and Snow had no choice but to wait patiently for a couple of days until a full list of all the goods on the trailer was received at the customs office. The officials unloaded everything, checked every item thoroughly, making sure that they all tallied with the documents, and then reloaded the trailer. Only when the officials were entirely satisfied that everything was in order were Scutts and Snow allowed to continue their homeward journey.

By mid-1969, Asian Transport trucks were leaving the UK on a more or less weekly schedule, heading to the customs compound in Avenue Takhte Jamshid, Tehran. The journey from door to door took about twelve days if the going was good. Loads included pharmaceuticals, household goods, electronic components, laboratory equipment, machinery spares and, of course, razor blades and whisky.

When the trucks arrived in Tehran they were held in the compound awaiting customs clearance and unloading. The drivers were not allowed to stay in the compound with their rigs so they stopped for a few days at their favourite hotel, the Continental in the centre of town. It was a grand building frequented by many British expatriates who were working in the area.

As the 1960s drew to an end, it had become obvious to British exporters that Asian Transport was the expert in overland transport to the Middle East. As far as Woodman and Paul were concerned, they had the monopoly on ultra-long hauls and that suited them perfectly. Everyone coped with the business very well but, as Paul said, what made things easy for them was that no other transport competitors were working from the UK:

'We were the only ones going overland from the UK to and from the Middle East. The only other companies were those based in Turkey and Iran, but their trucks and loads all terminated in Munich, Germany. They weren't allowed to come any further. There was absolutely no one doing complete round trips, door to door except for us.'

Paul admitted that in fact there was one other company which did set up and was very keen to muscle in on his work, although it was not engaged on faraway Middle Eastern destinations at the time. It was the Turkish Trading and Transport Company (TTT)

It was a hard life at times! Graham Wainwright (centre) and John Frost relax by the pool at the Continental hotel, Tehran. (GP)

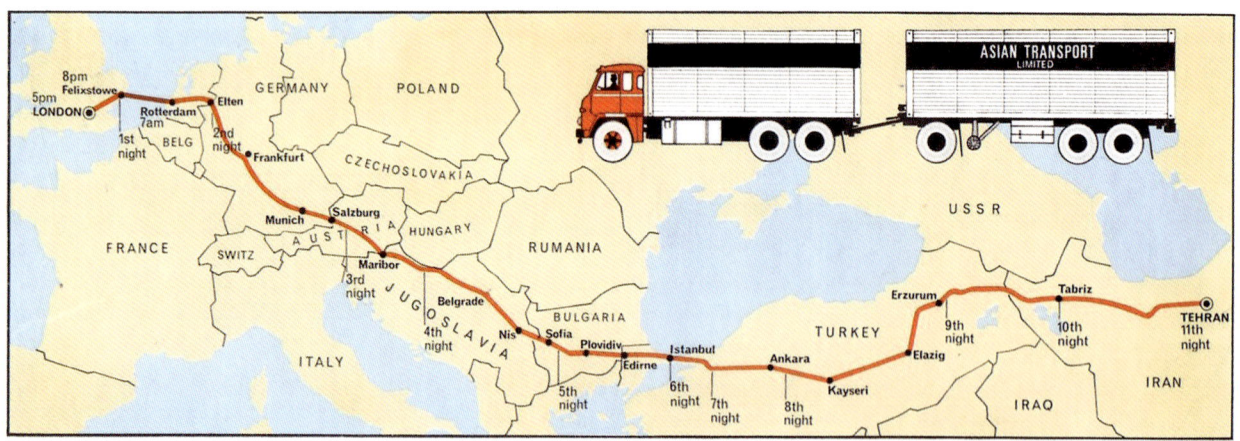

Publicity flyer showing the route and days taken from London to Tehran. (ASTC)

The Early Years

Turkish Trading & Transport Ltd (TTT) was owned by Mike Sherif who boasted that he would take over from Asian Transport. (RC)

owned by Mike Sherif whose family was of Turkish origin. Paul said that TTT was really the only threat to Asian Transport in the early days, purely because of Sherif's connections in Turkey.

Sherif had begun making plans to operate a transport company in 1955 when he investigated distributing fresh Turkish produce across Europe, but it wasn't until 1968 that he finally put a vehicle on the road and started his company. Luckily for Woodman and Paul, Sherif chose to concentrate on the refrigerated transport of perishable goods rather than the dry goods which were Asian Transport's speciality. In his first year Sherif transported over five hundred tons of grapes, melons, peaches, apples, oranges, tomatoes, artichokes and capsicums (chilies) to the wholesale markets of Britain.

Sherif's dream was to have a fleet of a dozen or so vehicles ferrying his native country's produce to the fruit and vegetable markets of Europe, but for some reason it never happened and his services only lasted a year or so. Bob Paul recalls the only time he met Sherif:

'I was sharing a truck with John Frost and we met Sherif at a restaurant somewhere in Turkey. He came over to us and started going on about this and that and I just sat quietly listening. He said to Frosty, "I know your boss, Bob Paul, and I'll be taking over from him soon, you mark my words."

'Then Frosty stood up, pointed to me and said to Sherif, "Meet my boss, Bob Paul …" He never said another word.'

By the end of 1969 Woodman had moved into an office above a greengrocer's shop in Chislehurst which Robert Gooda had found for him. As the workload for Asian Transport was continually growing and more vehicles and employees were joining the company, Paul reluctantly made the decision to come off the road and spend his time in the office looking after the ever-increasing mounds of paperwork and administration.

The office had a staff of two to help them run things. 'There was Mr Reece, and I always addressed him as that,' Paul said. 'He looked after the accounts. We also had Sharon who was our lovely secretary.'

Peter Cannon and John Frost shared AMY 147H and were tasked with taking the very first load to Islamabad in Pakistan. The strange looking vehicle, resting at a local village market, always attracted attention. (PC)

Such was Woodman's desire to push his company that he offered Tony Soameson promotion to sales manager. Soameson, who already had sales experience, was delighted by Woodman's offer. Until now he had been sharing a truck with Peter Cannon but relished the chance to come off the road to continue with his sales career.

With no depot or warehouse facilities, many of Asian Transport's outgoing loads were picked up at the warehouse of freight forwarder Baxter Hoare near Blackfriars Bridge in central London. They would consolidate everything and make it ready for the trucks. Once they were loaded, the vehicles would be driven to Chislehurst and parked in the railway station approach opposite the office, while the drivers waited for relevant paperwork and documents to be processed.

Once he had come off the road Paul began to miss life in the trucks greatly, especially when the drivers arrived at the office and he could see the trucks parked outside.

But he and Caroline had started a family so he wanted to spend more time at home. As he was no longer on the road, he decided to promote Gordon Pearce to road foreman to look after the other drivers while they were away. With his experience and calm nature, Pearce was ideally suited for the position.

Although Paul was no longer on the front line, he always kept in touch with his customers and at times did go back to do some relief driving. He recalls a conversation with Huffner of OCM about a particular load of carpets:

'Huffner used to phone me at home and say, "What the hell are you doing there, Paul? Why aren't you in that bloody truck? Your driver's late. I want your wife to be pregnant again so you can get back from Tehran in five days with my precious carpets."'

Until 1970 all of Asian Transport's traffic was destined

for Iran with the exception of a rare trip or two to Kuwait, but at the start of a new decade the company was ready to expand further into uncharted territory. Peter Cannon, accompanied by John Frost, took the first load to the British High Commission in Islamabad, Pakistan.

The rig was loaded with a consignment of personal effects and light fittings. The journey went well until the last section where the men had problems with the route to Islamabad as their road maps turned out to be inaccurate. Their drawbar outfit was only lightly loaded so the trailer roof was a good seven or eight inches higher than the truck roof due to the height of the dolly coupling. Cannon said that they came close to disaster at a tight-fitting bridge under a railway line. Of course it was not marked on their map:

'I was driving and was concerned that we wouldn't get under. I asked some locals if there was another way around but they said no, so I had no choice but to go for it as I could not turn the rig around. Frosty and I decided to let all twelve trailer tyres down in an attempt to lower the overall height.

'I then drove very slowly under the bridge. The trailer roof scraped on the metal girders as I went through and the suspension springs were pushed right down. The whole thing was making a hell of a row but I got the rig through and as the back of the trailer came out, it popped up like a cork out of a bottle as the springs relaxed. I managed to drive for a few yards until we could stop in a safe place to inspect the damage and re-inflate the tyres.'

Unfortunately the men came across another problem. As the trailer had pushed down on the springs, they in turn had pressed down on the wheel rims and the inner tubes had all moved out of position, making it impossible to find the valves.

Cannon and Frost had to strip every wheel to re-fit the tubes, then rebuild the wheels, put the tyres back on and re-inflate them all again using the truck's own air compressor which was a very slow process.

It was a very hot day. The road became busy with local trucks racing past, hooting and kicking up great clouds of dust. It took them a good few hours of hard work to get the trailer back up and running.

After completing the journey and unloading in Islamabad, the men realised that they were never going to get back under the bridge. This time the truck was completely empty and so it was even higher:

'We took advice from the British embassy in the city as to which way we should go. We were then given a ride in the consul's personal Austin Princess complete with a white-liveried chauffeur. He drove us to the Afghanistan embassy where we bought visas for the return journey.

'I remember that when we arrived there were scores of local people pushing and shoving in line for visas and the guards were holding them all back with pickaxe handles and Lee-Enfield rifles. Frosty and I looked at each other for a moment wondering whether we would have to wait in line too. However, our immaculately dressed chauffeur ushered us straight past everyone and right into the consul's office where we got priority treatment and were given two visas to enable us to travel back through Afghanistan.

'We also asked the consul which was the best way for us to take the truck to avoid the low bridge and were told to go via the Tarabela dam which was being constructed a few miles away on the river Indus. As we made our way to it, we had to drive through some tiny villages which again brought us problems and we wondered if we had been given duff information.

'There were some low wires hanging across the roads so we got some of the children to climb up and walk along the top of our trailer lifting the wires as we went under them. Thinking about it, that really wasn't the best thing to do! We also had to move all the barrows and baskets of fruit and vegetables which lined the very narrow roads before our huge rig could get past.

'Eventually we got to the dam but were stopped in our tracks by a burly security man who told us we were not permitted any further. We were very frustrated after struggling to get there but after some lively discussion the guard fetched a manager who just happened to be English. We got his permission to drive through the works area.

'We found that quite difficult because we had to negotiate all the construction equipment and workers who were totally oblivious to our strange truck meandering through. It was a case of following the dump trucks up and down and round and round until we found our way out.'

Cannon and Frost finally got back onto the correct route and found they had driven over eighty miles on their detour. The instructions given to them back at the embassy were spot on and they came out no more than a mile from the low bridge that had caused the tyre problems. Thankfully they arrived on the correct side of it!

In August 1970 Asian Transport put its first tractor unit on the road. It was a double-drive Scania 110 6 x 4

ELK 385J, a Scania 110, was the first tractor unit that Woodman and Paul bought. Still covered in factory wax it is already working hard. Note it is right-hand drive. (GW)

ELK 385J, now in full Asian Transport red-and-yellow livery. Pictured in convoy with a hired Avis Volvo parked close to Mount Ararat in eastern Turkey, en route to Tehran. (GW)

The Early Years

and it marked the turning point away from the traditional drawbar outfits. Graham Wainwright was the lucky driver to be handed the keys to the new Scania, ELK 385J:

'There were four or five of us in Istanbul and I got a message telling me to fly home immediately. I was assured that nothing was wrong and when I got off the plane in London, Bob Paul met me and drove me straight to Saunders Garage at Hemel Hempstead. There on the forecourt was a brand-new truck which Bob told me was mine. It looked strange because it was still in primer and covered in wax but because we were so busy I had to take it like it was for the time being.

'I did two trips to Budapest to see how it behaved. Then it was painted in our colours and I took it to the Middle East. I remember one thing about it, it pulled like a train.'

Apart from being a tractor unit, the only major difference between this truck and all the others in the fleet was that it was right-hand drive. Wainwright said that it was that truck which gave him a dislike for singer Petula Clark:

'The new truck came with an eight-track tape player which was marvellous because I'd never seen one before. Because of this I hadn't got any tapes of my own and had to play the one that came free with the player - Petula singing all her best tunes. So all I listened to for the trip to Budapest and back was Petula bloody Clark.'

Along with ELK 385J, another brand-new truck was also put into service later that year, but that one was a traditional drawbar outfit, a 14-litre Scania 140 6 x 4,

ELK 384J, the first Scania with a V8 engine to enter the fleet, has arrived at OCM's London warehouse loaded with Persian carpets. Scania 110, TDX 400K, can be seen parked behind. (ASTC)

129

ELK 384J. This was the first turbocharged V8-powered vehicle to go into the fleet.

Michael Woodman had inspected the actual truck at the London Motor Show in September 1970 where it was an exhibit. He also took a short test drive at the Scania demonstration area. He was so impressed with the specification and the powerful engine that he bought the truck there and then and had it delivered straight from the show.

The 140 with the V8 engine was introduced to the UK for the first time at the show and, although it had been available in Europe since 1969, Asian Transport was probably the very first customer to buy one. The 350 bhp engine was much more powerful than anything Woodman had used previously and he hoped this would cut down journey times across the arduous routes to the Middle East.

Ron Bell was given the new V8 rig from new, but he said that unfortunately it gave him nothing but trouble from day one:

'It was a nightmare and suffered some terrible engine problems. Number four and number eight pistons used to get partial seizures. I used to have to carry spare pistons and liners just in case.

'I remember breaking down on the top of Bolu, between Istanbul and Ankara, in the middle of a harsh winter and I had to crawl under the truck to drop the sump to get to the parts. I had to light fires underneath too, to try and keep things warm enough to work on. I could only work underneath for about twenty minutes because it was so cold. I think I was stuck for about four days before I got the engine going again.

'I started off again and had driven about ten miles further along the road when the crankshaft snapped. The extreme cold had affected that too. It all ended up with me getting towed all the way back to the UK.'

Because it was a new type of engine there were quite a few problems. Dave Poulton was also charged with the same truck for a couple of trips to Iran. 'It was a jinxed motor,' he said. 'Whatever could go wrong, did go wrong.'

Gordon Pearce poses in his Scania 110, BGH 172H, as he loads for Kuwait and Doha in November 1970. (ASTC)

Pearce (second from left) and Woodman (centre) discuss the inaugural route to Doha with Cable & Wireless managers. (ASTC)

Peter Cannon recalls a humorous story about the ill-fated truck:

'The truck was so unreliable that one evening we all took bets as to how many miles it would do before the first breakdown of the trip. John Holland was driving it at the time, so off he went to Felixstowe docks. Ten minutes later he returned saying it had broken down.

'As John himself had bet a silly distance of fifty miles or so, the rest of us didn't believe that he had actually broken down so soon, thinking he was only saying that to pick up the money. In fact he was correct and he had had a problem within 500 yards of the office.'

In 1970 another new contract saw one of the trucks go to Abu Dhabi for the first time where a civil engineering project was being overseen by a British company. In November the same year the first trip to Doha in the southern Gulf state of Qatar was undertaken on behalf of Cable and Wireless using another new rig, BGH 142H, driven by Gordon Pearce accompanied by Ray Neale.

Unknown to them, they were about to open up the most famous of the destinations to which Asian Transport/Astran has ever been in all its years of business. Pearce and Neale were the first Englishmen to take a commercial vehicle overland to Doha.

Like anything new, there are always one or two teething problems and that inaugural trip to Doha was no different. There were problems acquiring Saudi Arabian visas before the men set off from London and they were told by the Arabian embassy to buy them at the border. Unfortunately there were some diplomatic problems when they arrived at Rafha on the border between Iraq and Saudi Arabia and it was found to be impossible to buy any visas there. To get them would have taken too much time and been very expensive.

As it was the first trip through Saudi Arabia, there were no relationships built up with the officials of the country which may well have contributed to the problems. Pearce and Neale found a local agent through whom a Saudi Arabian national was employed to drive their truck across the country while Pearce and Neale flew to the other side to meet him at the border into Qatar.

Unloading at Doha, December 1970. Not a hard hat, hi-viz or safety harness in sight! (GP)

As the road journey would take at least two days, there was no rush for Pearce and Neale so when their plane made a scheduled stop at Bahrain, they took the opportunity to stay for the night and catch another flight the next day. They found a nice hotel which, according to Pearce, even sold decent bacon sandwiches.

The following day the men arrived at Dhahran on the Saudi/Qatar border to find that the rig had been left there by the Saudi driver who had then gone back to his base. Pearce and Neale still had to wait a few hours for their truck to clear customs. To compound the delays even further, they then found that the officials had stripped the cab out completely in a routine search.

When the authorities finally gave Pearce permission to retrieve his beloved truck he found the cab filled with thousands upon thousands of fruit flies which took an eternity to clear out. As a gesture of goodwill, the Saudi driver had kindly left a large bunch of bananas in the cab as a gift, but unfortunately the flies had found them first!

After finding the customs compound in Doha and unloading the cargo, Pearce and Neale then spent a couple of weeks scouting around looking for a return load.

At that time Doha was a small development but there were some British ex-pats living and working there. Pearce told me that he eventually managed to secure a small amount of personal effects destined for the UK. The men were also able to buy Saudi visas from the country's embassy in Doha which meant they had no trouble travelling through the country on the return leg.

The Asian Transport fleet of drawbar outfits had always proved very successful, their flexibility being the most important factor. The vehicles had been built to the same design for the type of cargoes which were regularly carried, However, with the emergence of new work and new destinations, Woodman and Paul thought they should invest in different types of trailers to keep up with the ever-changing and rigorous demands of their customers.

The original drawbar box-van trailers were suddenly becoming too small and restrictive. Even though Paul had previously hired different types for specific jobs, he and Woodman felt it was now time to buy their own new equipment.

They opted for a variety which included 'tilts', or canvas-covered trailers, which were particularly suited to top and side loading. They bought larger and longer steel-sided box vans designed primarily for consumer goods and household effects as well as low loaders which were perfect for handling indivisible and wide loads. Responsibility for buying all new equipment was shared between Woodman and Paul but, Paul said, 'Woodman was much better with the trailers.'

By the spring of 1971 Woodman had given up his dentistry completely in order to help Soameson attract even more business in the UK and the Middle East. Now he was more often than not away from the Chislehurst office. Being a man down, Paul decided to ask Peter Cannon if he would consider coming off the road for a managerial position.

After some thought, Cannon took up the offer and swapped his cab for a life behind a desk. Cannon is unsure of the exact date that he moved in, some time in late 1971 or even early 1972. 'We were living in Norfolk at the time,' he said. 'I had to commute weekly down to London and sleep in a spare room at the Chislehurst office for a year or so before making the big upheaval to move my family down to Kent for my bright new future.'

Not only did Cannon deal with the paperwork, he also had to visit various warehouses where the trucks were loading to make sure his documentation tied up correctly with the cargo:

'Loading at Baxter Hoare's warehouse, I had the manifests with everything on that I could load into the trucks. Asterisks marked the cargo which was urgent and had to go first. The idea was not to send out one truck fully loaded weight-wise but having little in it and the next one full to the gunnels but only minimal weight. You had to judge it so that both trucks were heavily loaded and if possible full cube-wise.

'That was no easy task as a typical groupage load consisted of 50-gallon drums, steel pipes, fragile cartons, weak crates and some awkwardly shaped elbow pipe joints. In fact it was a nightmare to get everything loaded sensibly. I remember always taking a lot of time and trouble over it as the load would be earning good revenue for a couple of weeks. I also remember that the warehouse staff would get irritable as they had their own work to get on with, apart from loading our rigs.

'When I first started in the office I used Bob's or Mike's Ford Cortinas to go up to Baxter's in London in the mornings and keep an eye on the loading. Then in the afternoons I'd race back to the office to continue with more paperwork, but that wasn't the end of it.

'The trucks would leave London around 4 pm and travel to Felixstowe docks to catch the night-time ferries. Meanwhile I was still typing away like mad to get all the relevant documents ready before they left. Then I'd jump in the car and race to Felixstowe to meet the drivers. If I was lucky I'd grab a cup of tea and a sandwich with them and then leave about 11.30 pm, drive back to Chislehurst, doss down in the office for a few hours and then start all over again the next day at 7 am in the London warehouse.'

As the errands became daily events, Cannon was presented with his own company car to make things easier: 'It was a Minivan. Being a tall chap, I had a hard job to get in - never mind sit in it comfortably. I got used to it and it did the job.'

During 1971-2 Woodman and Paul decided to put another three Scanias into the fleet. Two of them, JAN 774K and JLL 688K, were drawbar outfits fitted with big V8 engines, and the other was a tractor unit, TDX 400K, with the smaller 10 litre engine.

Bobby Vallas, who was given TDX from new, said it was a fantastic truck:

'It was double drive and had heavy-duty axles which made it very strong. It had a long wheelbase which made it perfect for towing the others out of trouble. It just gripped the roads so well.'

The ever-increasing volume of imported carpets meant that more and more space was needed. Woodman was hungry to utilise every inch so he designed his own trailer especially for OCM carpets. After much consultation with trailer manufacturer Crane Fruehauf, his original idea was deemed impractical, but with some patience and careful alterations, a very similar trailer was eventually manufactured and put on the road - much to Woodman's delight.

It is interesting to note why Woodman and Paul chose to use Crane Fruehauf trailers. In 1964 when the two men set off on the first trip to Afghanistan a York trailer was loaned to them on the understanding that the company motif and name would be splashed all over it. In return no rent would be charged.

After all three trips were completed Woodman

The Long Haul Pioneers

received a bill for the hire of the trailer. He and Paul then visited York's offices where a lively debate ensued, the outcome of which was that York insisted the money should be paid to them otherwise there would be a court case. Asian Transport had no chance against the much bigger York Trailer company and so the bill was settled grudgingly. This ended the relationship with York.

Paul said that in the 1980s when Astran was at its biggest there were about thirty trailers in the fleet, but none of them were manufactured by York.

Many of the trailers in the Asian Transport/Astran fleet were leased, mainly from Transport International Pool (TIP). Later on in the 1980s a batch of about twenty was bought from Gil Flex Rentals, all of which remained in the rental company's distinctive red livery.

The brand-new Scania 140 V8, JAN 774K, at rest in Germany during her inaugural trip. (ASTC)

Another brand-new Scania 140 V8, JLL 686K, parked outside the Chislehurst office. Note the use of both Asian Transport and Astran sign-writing during the re-branding period. (ASTC)

TDX 400K coupled to the custom-built trailer that Woodman designed. (GP)

Astran was operating about seventy-two trucks during its busiest period in the mid-1980s, most of them being sub-contractors with hired trailers. Peter Cannon said that the greatest number of trucks abroad on any one day was fifty-six.

Around the period of 1971-2 Woodman decided that the company's expansion would be helped if they re-branded it with a catchy name that sounded more modern. Paul described a further, unexpected, reason for the decision:

'Mike used to take the trucks for loading and he used to get stopped quite regularly by the police as he made his way in and around London and then to the docks. They wanted to know what Asian Transport was doing driving around leafy Kent and so on. It was all very annoying and time-consuming for us and we didn't want to attract any bad publicity either.'

It seemed that the boys in blue were suspicious of the company because of its name. Even though everyone knew it was a transport business, the police didn't see it that way and were not convinced by Woodman. So to stop any further enquiries, a simple but very effective solution was found. The two words of Asian and Transport were mixed together to form ASTRAN.

One of our bodies is missing! UHM 26F looking rather worn parked outside Graham Wainwright's house. (GW)

Flat spot. Gordon Pearce and Bob Paul, sharing UHM 26F, assist Ray Scutts and Dick Snow, sharing WLO 95G, with some puncture repairs and wheel changing. (RS)

Wot ... no trailer? Sometimes the drawbar trailers were left at customs compounds while the truck made deliveries elsewhere. The radio on the dashboard would usually be tuned into the BBC world service. (RS)

CHAPTER 6

Happy Days

Now firmly established as leaders in overland transport to the Middle East, Astran International, to use its full title, was in a strong position to keep up with the huge customer demand. It was decided that the newly branded company would be responsible for shipping, forwarding, groupage loads, insurance and packing. Asian Transport would be kept running alongside to continue with the overland road services. Providing the exceptionally good service continued as it always had, Astran couldn't put a foot wrong.

Until now Asian Transport had specialised in only single and larger consignment loads. However, some of Asian Transport's existing customers had begun to enquire about sending smaller consignments and part-loads to the Middle East, a demand which could now be fulfilled with the formation of Astran.

Due to various wars, particularly those between Egypt and Israel, which had caused the closure of the Suez Canal, many sea cargoes to countries surrounding the Persian Gulf were going via the Cape of Good Hope. This was expensive and slow. Michael Woodman firmly believed that this helped develop his business because his trucks could make the overland journeys much more cheaply and quickly.

There seemed to be no let-up in the lengthy shipping delays and so more and more shippers were turning to the driver-accompanied overland services. Astran continued to reap the rewards - much to Woodman's delight. With an abundance of work, he was able to buy more trucks and equipment. He also took on his first sub-contractor, John Cooper, during 1970.

In 1972 the turnover of Astran International and Asian Transport was over half a million pounds and the cargoes that the trucks carried were worth over six million pounds. Some of the export loads included vehicle spare parts, textiles, confectionary, machinery, furniture, electronic components, oilfield equipment, ship's spares, pharmaceuticals, whisky, the odd load of razor blades and even an armour-plated Rolls Royce.

The main cargoes for return loads to the UK were highly valuable Persian rugs, nuts, citrus fruits, personal effects, gum tragacanth and even a broken down helicopter coming to Britain for an overhaul. Between them, the Asian Transport and Astran trucks completed over two hundred trips and covered over 1¾ million miles.

By the end of the year, Astran had set up distribution centres in France, Germany, Holland, Italy, Czechoslovakia, Hungary and Yugoslavia, all of which were linked to the UK and the Middle East. To head up

When the company was re-branded Astran International the name Asian Transport was kept running alongside. (ASTC)

the European centres, Woodman arranged to have an office and handling warehouse in Holland registered as Astran International BV.

This was the start of the Middle East boom in overland road transport which lasted from the early 1970s well into the '80s. In 1974 crude oil prices rocketed following the Yom Kippur war of 1973, and this gave the oil-rich Gulf states unlimited capacity to spend, spend, spend! As many freight exporters had customers in the Middle East waiting impatiently, it is easy to understand why new transport companies suddenly began to appear on the Middle East run.

Sadly, many of those operators and owner drivers failed miserably. Some were only able to complete one or two trips, some never even got to their destinations and others had had enough by the time they got their loads delivered. David Miller summed up the problem well:

'In 1973 when I first made my hesitant way eastwards, there was probably a total of a hundred blokes, almost all British, who were involved in the job and I think it fair to say that there were very few pillocks amongst them. By the early 1980s that number had increased to perhaps five thousand of all nations and the pillock factor had grown exceptionally.'

Many of the new, inexperienced operators began sending their trucks to the same destinations as Astran, trying to poach work for lower rates of pay. Some of the new companies were properly set up and had the right equipment and employees, but the majority had no idea of what lay ahead of them, or the risks involved in such a specialist area of road transport. In the *Commercial Motor* journal of 5 October 1973 Michael Woodman said, 'They have looked at £15,000 contracts and gleefully rubbed their hands. However, because they did not cost the exercise properly, they finished up burning their fingers.'

Woodman realised that the good reputation he and Paul had built up was now in danger of being ruined by unscrupulous operators. He was aware that Astran now had serious competition. However, he knew that there was more money to be made from the rapidly expanding shipping and forwarding business, so he decided to concentrate more heavily on that, relying on Asian Transport's reputation.

As well as the overland driver-accompanied services and the new freight forwarding division, Astran was also asked to take large road-going vehicles direct to customers based in the Middle East. Besides being cheaper and quicker than trying to pack them into trailers, this was also a more practical method. In some cases it would have been impossible to unload the vehicles at their destinations, so driving them overland was the only option. When the deliveries had been made, the drivers flew back to the UK.

One of these jobs was given to Ron Bell who was tasked with delivering to Iran a specially built outside-broadcast television unit installed with expensive electrical equipment. The vehicle was a Bedford TK petrol-engined truck which also towed its own generator trailer. As Bell said, the journey went well until he arrived in the hot Middle East:

'It was a nightmare trip which took eleven weeks and it was all due to the Shah of Persia (Iran) having an argument with the ruler of Bandar Abbas and cutting his television off. The Crown Agents secured the contract to supply the TV unit and I got the job of driving it for Astran.

'Once I got through Europe and into Turkey, the radiator kept overheating every time the truck climbed steep hills. The engine wasn't powerful enough for the weight of all the equipment and the genny on tow. I had to keep stopping to let things cool down and refill the radiator. I had gallons of water on board which also made the vehicle heavy.

'Then when I got into Iran I had to make room for a policeman who accompanied me to keep an eye on all the expensive equipment as the vehicle had been sold to a wealthy Iranian and everything had to be in tip-top condition. It was bad enough having to live in the cramped cab, but now I had a passenger it got worse. We had to squash up when we slept at night in the back with all the equipment.

'As we travelled across the hot desert, the radiator and pipes kept bursting and it used to keep boiling dry. I had worked on diesel engines before, but never petrol ones. I had to learn how to fix it as I went along. When the radiator finally split, I had to take it off and then look for a local blacksmith to weld it up again. To make things even worse, the cab did not tilt and the only way into the engine was through two tiny little flaps on each side of the cab.'

After many running repairs, Bell had no choice but to give up when the engine finally seized. It was simply not made for the incredibly long journey and especially not the intense heat of the desert. The bodywork completely covered all the areas where air should have

Ron Bell had the company of this policeman once he entered Iran. (RB)

The Bedford cab did not tilt, so Bell had to work in a very confined space. The water on the floor is from the radiator. (RB)

been able to get in. At one point, Bell had no choice but to remove the side flaps to allow in as much air as possible, but even that failed.

Bell sought the help of Harry Sully at British Leyland in Tehran. Sully decided to send a recovery truck and tow the stricken Bedford back to his garage. Only 250 kilometres from his destination, Bell was desperately disappointed that he couldn't make the short last stretch of his journey.

Bell was picked up and towed 300 kilometres to Sully - back in the direction he had just come from. A new engine was flown out from England and while he waited for the repairs to be completed he stayed for about four weeks in the Mar-Mar Hotel in the centre of Tehran. 'It was a beautiful place,' he said, 'but I got fed up sitting in the bar all the time.' Of course you did, Ron!

With a new engine fitted, Bell and his policeman escort finally completed the journey to Bandar Abbas. Once the vehicle was delivered to the customs compound, the policeman caught a bus all the way back to the Turkish border while Bell enjoyed a flight back to the UK. 'After all that trouble, I never wanted to see another Bedford truck again,' Bell concluded.

Such was the demand for Astran's services that Bob Paul had to hire extra trucks. He chose Avis Truck Rental based in Euston, central London, as the supplier of some Volvo F88 6 x 2 tractor units. One of the trucks was painted in Asian Transport's distinctive red-and-yellow colour scheme after Paul agreed to take it on a long-term lease. The others were left in the standard Avis red livery, but did have the Asian Transport/Astran name in the headboard and on the cab doors.

Paul cannot recall exactly how many Volvos he hired. 'I think we had five or six as the most,' he said. 'But sometimes we needed just two or three as and when.'

Paul remembers being offered huge amounts of Green Shield stamps when he signed up to rent the trucks:

'I told the salesman I didn't want stamps so could he knock off the amount in money from the bill. He said it wasn't possible which annoyed me because I just didn't want the bloody stamps. The guy who ran the Avis depot kept pestering me to collect my stamps; he was always phoning me up saying I had millions of them waiting in his office.

'Eventually I went in and he handed over some huge envelopes stuffed full with stamps. Mike Woodman and I soon became known as the Green Shield Stamp families because we had all manner of things at home which we'd

Bun Parlane was the regular driver of JLB 279K, the only hired Volvo to be painted red and yellow. Pictured here in convoy with JLB 259K, the pair are at rest in eastern Turkey. (BPA)

acquired with them like washing machines and fridges, and Mike even had a fully kitted-out tent.'

From day one, all of Asian Transport/Astran's cargoes had been consolidated at various warehouses and customs houses in and around London with the whole operation being controlled from the office in Chislehurst. In late 1973 or early 1974 Woodman and Paul made the decision to move into their own depot. Paul wanted a location close to Dover to make it an easy journey for his trucks which were now mainly using that port for shipping in and out.

With the help of his sales manager, Tony Soameson, Paul found a suitable site to rent next to the A20 at Addington. He proudly named it the Middle East Freight Terminal. It had a purpose-built 48,000 square foot covered warehouse, a large garage and plenty of outside space for the trucks and trailers. There was even a small bungalow which served as an office and was perfectly positioned at the entrance to the depot so that the staff could keep an eye on everything.

Paul also employed Tony Keirle, who was sadly killed in a road accident a few years ago, as chief mechanic. He and two assistants, one of whom was Mick Payne, were taken on full time to service and repair all the trucks and trailers in the fully kitted-out workshops.

Keirle was an extraordinary mechanic for whom Bob Paul had a huge amount of respect. 'There was nothing Tony didn't know about trucks, especially Scanias,' said Paul. 'He was a godsend in my garage.'

Keirle had previously worked for Turkish Transport & Trading travelling around to look after the company trucks. He would drive – or fly if they were too far away. Then Peter Cannon told Paul that Keirle had set himself up in Istanbul with a mobile workshop to make himself more versatile.

Cannon explained how he knew about Keirle:

'There was an old generator engine in the basement of the BP Mocamp truckstop in Istanbul. Somebody had stripped it down and left all the parts spread out in a heap all over the floor. The owner had installed a new engine because nobody fancied the job of trying to put together a heap of bits that might or might not be complete. As Tony Keirle was based there, he said he would do his best to put the old genny back together in his spare time for something to do.

'The job took quite a while because Tony would be called out to anywhere in the country for a breakdown. Eventually he did complete the rebuild and had the satisfaction of seeing it running.'

Bob Paul thought that Tony would be a good bloke to have working for him: 'I kept the idea in the back of my head. Then some time later I heard he was working just up the road from us at the local Scania dealership so I went and poached him.'

Parked at Felixstowe, Bun Parlane and Brian Parkinson, just visible on the left, attend to some running repairs while they wait for the ferry. (BPA)

The Middle East Freight Terminal, Addington Kent. HKJ 540N, owned by sub-contractor Pat Riley, prepares to leave closely followed by TDX 400K. In the background is a regular French subbie. (TS)

141

After the opening of the new depot, the Chislehurst office was kept on for a time to serve as the accounts department and Woodman's main base. With the new set-up at Addington, the office was manned seven days a week to look out for any cable messages, telexes or take telephone calls.

This was an instant improvement. All the drivers, including the subbies, were asked to contact the office every three days or so by the quickest means. This meant that customers and families could be kept up to date with the progress of their loads and loved ones. There was always someone at hand if any of the agents based at border crossings or faraway cities needed help.

Transport manager Peter Cannon was very much at home in his new office at Addington. (PC)

Unknown to anyone when they moved into Addington, the icing on the depot cake was to be found just one and a half miles away in the village of Offham. The Kings Arms soon became very well known to all of

Inside the workshops. Mick Payne can be seen working on TDX 400K while Tony Keirle inspects the chassis of JAN 774K. (ASTC)

Astran's staff and drivers. Most lunchtimes and evenings there would be a mass exodus straight to the pub and, according to Peter Cannon, it wasn't easy to get out of going. It became a ritual but, as he said, it was better than sitting at his desk trying to answer the telephone while eating a sandwich.

When they returned from an arduous long-haul trip some of the drivers would want a party before their wives collected them. It was not unusual to find two or three of their trucks parked in the tiny pub car park at night. Dick Snow and Bobby Vallas certainly enjoyed a drink or two and they would stay in the pub until long after closing and then simply stagger back to their cabs and surface some time the following day.

When any of the wives rang the office to enquire where their husbands were, they would be told the men were 'cashing up' and that it might take them a while to get finished. Cashing up was a phrase used for exchanging foreign currency, sorting out paperwork after a trip and being de-briefed. In reality, drivers would more than likely be in the pub enjoying a well-deserved beer or two.

Sadly there was some dispute over an unpaid bill at the Kings Arms, so another pub had to be found to keep the tradition going. Everyone moved down the road to the Harrow at Ightham where they were made to feel extremely welcome. As the landlord got to know everyone, he would leave the pub at closing time and trust them to serve themselves, putting money in the till or writing their drinks on the blackboard if they only had foreign currency.

It was also arranged for the pub key to be left by the back door and some drivers would stay in and not lock up until 7.00 am the next morning. They were indeed 'happy days'.

As described in the previous chapter, right at the start Woodman had to abandon his aim to use only British-built vehicles. During the build-up of Astran, the majority of the Scania trucks were reliable and worked extremely well, but Woodman still wanted to use a British-built truck. It was just unfortunate that there were none available which could match the specification, performance and reliability of the Scanias.

The hired Volvos from Avis served their time well, but Woodman was simply not a dedicated foreign vehicle operator and so persevered in his quest to find some suitable British machines. In the *Commercial Motor* journal of 5 October 1973 Woodman is quoted as saying, 'I see that at last British Leyland has produced a vehicle to challenge the continentals. I would dearly love to buy British and I sincerely hope that the Marathon can stand up to our work.'

Woodman secured the use of a three-year-old Leyland Marathon and a Scammell Crusader for a trial period. Although neither was built exactly how he wanted them, both trucks were sent to Tehran and back to evaluate them.

The Marathon, WTJ 120L, was driven by Dick Rivers, employed by British Leyland as a senior demonstration driver, accompanied by Astran driver Jeff Ruggins. The rig was fully loaded to 36 tons and the pair started their journey at the end of November 1974. They reported that the weather was exceptionally bad on the outward journey once they had got into Turkey, but nevertheless the Marathon coped reasonably well.

When the men arrived in Tehran, the rig was taken to the British Leyland garage where some modifications were made to the front wheel arches and the rear metal ones were completely removed. The return trip to the UK took twelve days including a three-day stop to help fellow Astran driver John Williams reclaim his truck from a ditch in Turkey. (See Chapter 14.)

The truck which Williams put into the ditch, JLL 686K, can be seen in the photograph on the next page. Williams started out on his trip shortly after Rivers and Ruggins had departed on theirs.

The Scammell Crusader, MUR 501H, was driven by Ron Bell, accompanied to Tehran by an engineer from Scammell Motors who carried all manner of equipment to monitor the truck on its journey. Unfortunately the Scammell only had a day cab so, as there was insufficient room for two men, luggage, supplies and the engineer's equipment, a roof rack had to be fabricated. With no room to sleep in the truck, Bell and the engineer had to find a bed-and-breakfast every night which added a considerable amount of extra time on the outward journey.

Bell felt that the truck was not up to the journey at all:

'Sorry to be so blunt but it really was rubbish. The truck had a huge engine hump taking up half the cab so there was no room to move or even stretch your left leg out. The worst thing was it had no engine brake so it was a real nightmare

WTJ 120L, lent by British Leyland, was used on a trial run to Iran. Seen alongside JLL 686K, it is preparing to leave Addington in November 1974. (ASTC)

to keep it in control on the mountain roads. The brakes used to get extremely hot which worried me a lot so I had to keep stopping to check them and let them cool down.

'The truck would over-run in every gear because of the heavy weight of the load pushing it along which in turn didn't do the engine or the drive axles any good at all as the revs were always up in the red. It also kept breaking down which didn't do the manufacturer's reputation any good either – so by the time I got to Tehran I was literally limping along. It was a case of getting it into the Leyland garage to be fixed as quickly as possible.

'The engineer wasn't happy with the way his product behaved, but worse than that he hated the vast amounts of dust after we got into Turkey and the many potholed roads caused him a lot of discomfort as he was perched on the unsprung passenger seat. Funnily enough, he decided to fly home.'

In the Tehran garage the Crusader was stripped down to the extent of fitting new drive axles from an Albion truck in an attempt to improve performance on the return trip. This seemed to work: 'The Scammell went much better on the return journey with the different axles fitted,' Bell concluded.

In 1974 Astran successfully transported its first heavy load the 4,100 miles to Damman in Saudi Arabia. This enormous 31-ton electrical transformer was delivered in twenty-three days on behalf of GEC. The transformer formed only a small part of a huge order and the record load coincided with a time when Astran's trailer movements reached their highest level since the start of Asian Transport in the mid-1960s.

In June 1974 business was up by a hundred per cent on the same month in the previous year. This represented, in that one month alone, export trade worth £750,000 to the Middle East. Such was the demand for Astran's services that extra destinations were added to the already busy Middle East services. They were: Amman in Jordon; Jeddah in Saudi Arabia; Dubai and Sharjah in the United Arab Emirates. Two other new destinations of Beirut in Lebanon and Damascus in Syria were offered as en-route stops.

In an attempt to attract more import business back to the UK from overseas customers, sales and commercial manager Tony Soameson was asked to head up an

Another trial vehicle, this Scammell Crusader, MUR 501H, was not built to Woodman's demanding spec. It had no sleeper compartment, no engine brake and was unreliable. (ASTC)

The heaviest load carried to date: a 31 ton electrical transformer hauled by sub-contractor J & T. (TS)

GUL 599N and two other Astran trucks unloading at the Tehran Motor Fair in the mid-1970s. (TS)

exhibition at the Tehran Trade Fair in 1974. Astran had a stand perfectly situated in the middle of the show, complete with a huge white board containing the company logo and a bold route map from the UK to Iran. It also displayed all the other Middle Eastern destinations available.

Each city on the route was depicted by a flashing light bulb and the display also had photographs of the Asian Transport/Astran trucks, showing the different types of trailers and equipment available. It was all purposely designed to attract Iranian manufacturers and exporters to use Astran's overland services.

By today's standards of ultra-modern graphic displays, the trade stand description sounds crude and cheesy, but by all accounts it worked well and promoted Astran International as a professional company. In fact Soameson proudly boasted that his flashing light display attracted the Duke of Kent who was visiting the fair.

Because the overland trucks spent so much time out of the UK, they were all taxed privately and were therefore not allowed to work for hire and reward in the UK itself. It meant they could not tow a trailer on any UK public roads but were able to run solo to and from the docks.

As transport director, Bob Paul had done his homework and spoken to the Driver Vehicle Licensing Authority (DVLA) at Swansea who gave him the all-clear to run his overland tractor units on private tax. 'After all,' he said, 'they were out of the country for ninety per cent of the year, so why should they be heavily taxed for work in the UK?'

To overcome the problem of getting trailers from the docks to the depot and more importantly to customers, Paul bought a tractor unit specifically for work in the UK. Known as 'the shunter', it was a second-hand Volvo F86, HMK 501K, which came into the fleet some time in the mid-1970s and was originally driven by Tommy Thomas. As the trailers were more often than not loaded to their maximum weight of 23 tons, the shunter needed to be as light as possible to keep overall weight within the UK limit of 32 tons, hence it was so small. It had a low-powered 198 bhp engine - but by all accounts it seemed to cope with its duties pretty well.

The overland trucks would come off the ferries in Dover, drop their trailers in the dock area and return to the yard solo. When the trailers had cleared customs procedures the shunter would collect them and deliver the cargoes. It would then reload them and take them back to Dover docks ready for the overland trucks to take out again.

HMK 501K, the Volvo F86 shunter, leaves Dover docks with a heavy overland trailer. (SL)

It was quite common to see the little shunter towing a heavily laden trailer along the A2 towards Dover, closely followed by a solo tractor unit. Paul also had a list of local hauliers which he could call upon to help move the trailers should the need arise.

By the end of 1974 the fleet stood at about thirteen own-account vehicles. Around eight were tractor units and the others were drawbar outfits. There were also five or six units hired from Avis. The trailers were split between Crane Fruehauf and Merriworth, some of which were rented from various companies including Transport International Pool (TIP) and Transport International Rentals (TIR).

As new Scania trucks were put into service, the older ones were usually part-exchanged with the dealer. As far as the two 'pink ladies', UHM 25F and UHM 26F, are concerned it is unclear when they were finally taken out of the fleet. Bob Paul does remember part-exchanging 26F which Ron Bell took on its final journey to the Scania dealership in Essex where he swapped it for a brand-new Scania 110, WLO 95G.

Bell said that as far as he knew 26F sat in the dealer's yard for many months and every time he went past, the faithful old Vabis looked more and more tired.

It is also unclear what happened to the sister Scania Vabis, UHM 25F. Dave Poulton was fairly sure that it was abandoned at Gradina on the Yugoslavian/Bulgarian border. Apparently the poor old truck sat there for many months after someone had dumped it.

When the road at the border was modernised and rebuilt, the workers buried it where it stood. If that story is true, then it's another sad ending for yet another truck which had such a rich life.

Poulton recalled the time that Bob Paul took him to the Scania dealership to collect a brand-new truck:

'It was the Scania 140 V8 double-drive, PHK 172M, tractor unit which cost about £8,800. Bob told me it was for me and, as you can imagine, I was as pleased as Punch. It was the best truck I ever drove for Astran and it pulled like a steam train, but I have to tell you that it did break down a couple of times.

'The best example was when the fuel pump broke. I was parked on the side of the road in Saudi Arabia almost in Qatar and the police arrived saying they would send someone to tow me to a garage. Soon after, a local truck arrived with a long steel bar which we attached between us and off we went. It took ages for the driver to get going as I was quite heavy and he only had a small engine.

A rare sight - five of the fleet at home all at once. Left to right: GUL 588N, NTW 562M, JAN 774K, AMY 147H, VGF 897M. (PC)

'The road wasn't the smoothest either which didn't help and as we were going down a bit of a hill the towbar just rattled itself off my front bumper. I'd already taken the prop shaft off so had no drive at all and I'd drained all the air out of the trailer so I had no brakes.

'My trailer was loaded and as the momentum kicked in I had to overtake the bloke who was supposed to be towing me, otherwise I would have smashed into the back of him. I swerved my truck out and found myself going faster and faster. As I went past his cab, I looked across and gave him a toot and a wave. His face was an absolute picture.

'I must have been doing about 70 mph and I have to admit I was worried. Luckily there was an uphill section which slowed me down. As my truck came to a standstill, I had just enough grip on the dead man's brake to hold everything until the other bloke arrived with some wheel chocks.'

It wasn't only drivers who were regularly joining Astran; as business boomed more administration staff were also required. On 21 April 1975 Jan Payne began the employment with Astran that is still continuing in 2009. Payne began as a copy typist and took up her position in the roof of the bungalow at Addington. Squashed in with five other girls, it was a somewhat claustrophobic environment to say the least. After a time she moved onto more general duties, including load documentation and dealing with Bob Paul's duties when he was away.

Although drivers were not permitted upstairs in the bungalow, occasionally one would creep up to say hello to the girls. According to Payne:

'Dick Snow was famous for doing it. But you couldn't really argue with him, could you. He was quite frightening at times with his imposing stature. I think I blotted my copybook with him once when I wrongly transposed a number on his paperwork. "I could have ended up in jail!" he shouted at me, to which I replied, "But you didn't, did you. You're here. So it couldn't have been that bad."'

Luckily for Payne, she was moved out of the bungalow into better accommodation when some metal Portakabins were erected further up the yard:

'To say it was better wasn't really the truth. You would boil in summer and freeze in winter. So that didn't last too long and we got moved back into the bungalow after it had been extended, but this time we were downstairs next to the sales office in a much nicer place.'

Payne recalls her working environment being very friendly especially at lunchtimes when everything closed down: 'It was my favourite time,' she said. 'We all went to the pub. I'd never known anything like it

Happy Days

PHK 172M cost about £8,800 in 1974. Dave Poulton drove it from new. (DP)

before but it was what made Astran such a great place to work. Everyone was so laid back.'

The summer of 1975 saw three brand-new Scania 140 V8s delivered: KJN 671P, KJN 680P and KVX 859P. Although Bob Paul was predominantly running tractor units by now, the new vehicles came as drawbar outfits which were, in theory, better suited for the groupage loads comprising many small consignments for different customers. Previously these types of rigs were specified as 6 x 2 or 6 x 4 trucks, towing two- or three-axle trailers, but the three new ones all came as 4 x 2 trucks towing three-axle trailers.

Unfortunately, this set-up caused weight distribution problems when the rigs were fully loaded because the trailers could carry more weight than the trucks due to their having more axles. This made driving very difficult at times, especially during the harsh winter months when the heavier trailers would push the lighter trucks into a compromising position on dangerous road surfaces.

Even though the new Scanias were fitted with the latest powerful engines, they were only single-driven axles, so they often suffered traction problems – especially in the mountainous regions of Turkey.

It is unclear why the new rigs – which proved unpopular with the drivers - were built as 4 x 2s and not to Woodman's tried and tested ways of the early days. Unlike the tractor units, the new drawbar outfits couldn't drop their bodies in the docks and travel solo, so they had to bear expensive road tax, much to Bob Paul's dislike.

KVX 859P was one of three new road trains which were not successful. All three were eventually cut down to tractor units. (BR)

To make matters worse, the vehicles were more often than not carrying a mixed load of groupage with more than one delivery in the UK. This consumed a lot of precious time as the delivery points were not always in the same area.

To try and speed the process up, the drawbar trailers were unhitched from the trucks and a standard tractor unit deployed to unload and reload them. This caused yet another problem because Woodman had specially designed the trailers only to be used with their own dolly couplings onto the drawbar trucks. When the standard units coupled up, they had difficulty turning, which often resulted in wheel arches and lights being ripped off.

Peter Cannon tried on numerous occasions to persuade Woodman to get rid of all the drawbar outfits and standardise the fleet with tractor units which he said would be so much easier to manage. Cannon was also convinced that they would be just as profitable as the drawbars. Woodman was used to his methods of loading the drawbar outfits to utilise maximum cubic capacity, so he was not keen to change his fleet completely.

However, he agreed to spend a couple of weekends at home with his calculator going through all the paperwork – only to find that Cannon did have a point and there was no real difference in the revenue between the drawbar outfits and the artics. If an acceptable mix of groupage was loaded, a good amount of money could be made. Cannon persevered with his idea of getting rid of the drawbars and eventually, much to his delight, he was heard.

Rather have the newest three drawbar outfits sold, the trucks were cut down and converted to standard tractor units. Their purpose-built drawbar trailers were modified slightly so that standard tractor units could

KVX 859P after she was cut down and repainted. Pictured here is Geoff Frost. (ASTC)

pull them properly. They were then specifically used for hauling event/show equipment and the personal effects of UK citizens going to and from the Middle East.

One of the trailers was also put to very good use at Christmas when it was placed next to the front door of the bungalow at Addington. The company party was held in the building with the disco set up in the trailer.

As well as being well known for using only Scania trucks, the Astran fleet had become famous through Europe and the Middle East for the very distinctive red-and-yellow livery which they had been using since the late 1960s. The bold letter A which represented the name Astran and also depicted a road, was painted on the bodies and trailers, and could easily be spotted from great distances.

However, as famous the red-and-yellow livery was, Woodman decided it was time for a fresh look for his company. In late 1975 he employed a consultancy which came up with various new colours and designs to complement Woodman's original idea of the big A. The chosen livery was a new bronze-and-white scheme, officially known as 'ochre and white'. However, it was never really accepted by the drivers or staff and soon earned itself the nickname of 'ivory and shit'!

Above all, the trucks suddenly became quite plain compared to the striking red-and-yellow colours. Nevertheless, the ochre-and-white livery was applied to the latest batch of new trucks and also to some of the older vehicles to keep the fleet uniform. The new livery lasted for three or four years until it was finally agreed that perhaps the famous red and yellow should be brought back.

Thinking about the most attractive truck that Astran ever bought, a case could be made for NVW 484P, a Scania 140 V8 double-drive tractor unit which entered

One of the road train trailers was put to very good use at Christmas. The party disco was housed inside it! The lighting was run from a 24 volt supply. Note the red and white trailer marker lamps hung on the wall for decorations. (AM)

NVW 484P looking resplendent in the new Astran livery of ochre and white against the desert backdrop. (DP)

The ochre-and-white livery was not liked by the staff. John Frost secures some luggage to the roof rack. (ASTC)

the fleet in the spring of 1976. Looking resplendent in the new livery and coming complete with all the bells and whistles, the truck looked stunning coupled to its regular box-van trailer at home in the desert.

Dave Poulton was the lucky driver who had NVW 484P from new, but unfortunately his enthusiasm for the truck was dampened by its bad braking problems which took an eternity to resolve.

Poulton had complained bitterly from day one to chief mechanic Tony Keirle saying that he couldn't stop the truck properly and was desperate for something to be done before it killed him! Poulton explained the problem:

'After the first trip to the Middle East, Tony completely re-lined the brakes on both the tractor unit and the trailer that I normally pulled but it still never braked properly.

There was nothing else he could do, so I just put up with it and got to know how to drive the rig and compensate for the bad brakes.

'Then when Tony took it for its first MOT, it failed because there was no proper air pressure getting to the braking system. None of us could understand why. Tony was an excellent mechanic who knew everything there was to know about Scania trucks, but this problem baffled even him.'

After numerous attempts to sort the brakes out, Keirle threw in the towel and contacted the truck manufacturer's head office in Sweden. It eventually transpired that NVW 484P should never have been sold as a tractor unit. It had been specially built as a rigid for a cement mixer body, so exactly how it ended up as a tractor unit is anyone's guess.

Once the cause of problem was revealed, the truck

UAR 747S, driven by John Bruce, and PWM 116T, driven by sub-contractor 'Rita' Haywood, en route to Doha in the late 1970s. By then, the red-and-yellow livery was being re-introduced. (JB)

was returned to the workshops where all the brake valves and pipes were changed for the correct-sized ones. From then on, the truck performed like it should have done - much to Poulton's relief.

One of the men helping Keirle in the workshops was Mick Payne, Jan Payne's husband. He began working for Astran at the same time as his wife and stayed with the company until 1979. 'I spotted an advertisement in the local newspaper,' Payne said:

'Having served my time at a Leyland retail garage and reaching technician status, I got bored, as you do, and decided to widen my knowledge of vehicle repairs. Astran gave me the ideal opportunity to do so in the 1980s with a fleet consisting of Scanias and then Mercedes and a sub-contractor's Volvo F89. There were step-frame box trailers, 40 ft tilts and drawbar outfits to work on, and every conceivable repair was kept in house so the spectrum of work was far more interesting than that of a retail garage.

'The vehicles got quite a bashing on the Middle East run so when they came back a bit of an overhaul to each was inevitable. It was more than just dropping out the oil and changing the filters. We sometimes had to change the cylinder liner O rings as they hardened during the trips. Scania provided weep holes in the side of the engine block. They were positioned between the two lower O rings so when oil was seen weeping out, it gave us time to change them before the oil got into the water system.

'Gearbox range changes took a bashing too, with the synchromesh rings wearing out through sheer usage and high temperatures. It would have made gear changing quite difficult and noisy. A common fault with the gearboxes was the lock nut coming loose in the end of the main shaft, thus making gear synchronising painful. Not good when you're thousands of miles away from base.

'The trucks and trailers were turned round within two weeks to a high degree of repair. It was not unusual to work right through until three o'clock in the morning if we knew a driver was booked to catch a ferry that same day. That was great fun during the winter months.'

The Astran fleet was reliant on Scania trucks but there were one or two infiltrators during the 1980s. Which marques did Payne prefer to work on?

'I did prefer the Scanias to the Mercedes because I felt the

engineering was of better quality. However the P-registered Scanias bought just after I started had recurring faults with the chassis cracking just behind the cab ram pivot point. We ended up welding and reinforcing the area as it would have been too costly for Scania to put right. It did the job until they were sold on.

'John Holland drove a Scania 110 and was always having problems with the diff trunion bolts which kept working themselves loose. I heli-coiled the diff and fitted new bolts with lock wire which lasted two trips. I ended up welding the trunion to the diff, new heli-coils and bolts and, hey presto! it lasted the life of the truck.

'If Dave Poulton's truck, NVW 484P, was coupled to his normal box trailer, T11B, the rig was too high to drive under a particular bridge on the Khyber Pass in Pakistan, so I fabricated and welded a hook onto the rear of his truck chassis which was in a lower position than the conventional fifth-wheel coupling. He was able to hook the trailer to that point and, with a safety chain in place, he was just about low enough to negotiate the bridge. It was a bit dodgy, but it worked.'

Payne has fond memories of the drivers. They were all efficient at running repairs, but one or two were not as house-proud as perhaps they should have been:

'I remember on more than one occasion, we had to pour Jeyes fluid over someone's air-conditioning radiator and then run the system for about an hour to clear it. The stench was so bad we couldn't do any work in the cab until it had cleared. I'm not going to mention the driver's name, but we had to do this on his return from every trip.

'Some of the drivers were more careful with their trucks than others. John Williams was probably the most mechanically minded and hands-on driver we had. He was forever poking his nose into things and by-passing every electrical fault as it happened, which was a nightmare for us when he got back - but he did always get back.

'He used to drive Tony Keirle up the wall with his make-shift repairs. But I have to say if you were stuck anywhere, Williams would be the one to get you back on the road and, more importantly, back home. A bit like Tony, I suppose.'

John Williams in the workshop at Addington. He used to drive Tony Keirle up the wall with his makeshift repairs. (UNK)

The convoy loaded for Afghanistan and ready for the off, GUL 582N in the foreground. Parked behind, left to right: NTW 562M, GUL 588N, TDX 400K and an uknown sub-contractor's Volvo F89. (ASTC)

Near the end of 1975, John Frost was part of a convoy with five other Astran trucks to deliver six Hymac excavators worth £100,000 on low-loader trailers to Pul-i-Khumri in northern Afghanistan, close to the Russian border. It was the furthest north that Astran had ever been. It was also a first for the excavator company as they had never exported any equipment to Afghanistan before. The 4,800 mile journey from south Wales took about twenty-one days with the machines finally being off-loaded at a new irrigation system.

1976 saw Astran International clock up 18,000 tons of freight delivered to the Middle East in the first ten years of business. The trucks had travelled approximately fourteen million miles in that period!

At this time, Iran was still a very important market but when the Shah was deposed, import freight to the country all but dried up. The Astran trucks were then diverted to the southern Gulf states whose construction and oil industries were beginning to develop rapidly. Qatar soon began taking the lion's share of the cargoes.

Journey's end. NTW 562N and GUL 588N arrive at the Salang Pass, northern Afghanistan. (JF)

In 1976 John Frost successfully delivered these 47 foot long horse racing stalls to the Arab Games. (ASTC)

In 1976 during the lead up to the UK's summer holiday factory closures, Astran was inundated with requests to transport freight to the Arabs. Sales manager Tony Soameson said in the 13 August 1976 edition of *Freight News Weekly* that:

'So far we have managed to cope with the demand, but if it continues like this we may be forced to turn business away. As international forwarders we use carefully selected outside contractors but there are just not the right vehicles or drivers available in the quantities needed to meet the current demand.'

All the Astran trucks were fully booked with most engaged on a contract delivering PVC cables to Qatar for the new telephone network. The trucks were leaving the UK at a rate of two vehicles a week, each trailer carrying 7,000 metres of ducting.

Having gained experience of abnormal loads with the Hymac excavators, John Frost was tasked to deliver a 47 ft one-piece section of horse racing starting stalls to Qatar's ruler. The journey was trouble-free through Europe, but Frost encountered plenty of hassle as he tried to negotiate the hairpin bends on the Turkish mountains.

Nevertheless, in just thirteen days he delivered the stalls in one piece, just in time for the Arab games of 1976. He had cut at least four weeks off the comparable sea transit time.

The wealth in the southern Gulf states was now growing at an extraordinary rate due to the construction and oil businesses. David Miller, who spent much time in Qatar, comments:

'When I first went to Doha, the capital of Qatar, in 1975, it was the size of a small English market town clustered around the port with a few modern buildings out towards the airport.

'Then they began to build ring roads round the town starting with A, which was just outside town, and ending with G that was so far out in the desert that you could no longer see the town. Then they built a stadium on ring road G for the 1976 Arab games. By the time I last went there in 1981, the stadium was within the town.'

The rich Arabs had more money than they ever dreamed of but no idea of what to spend it on. With their new-found wealth, they wanted the best of everything. Miller continues:

'There were brand-new top of the range cars like Rolls Royce, Jaguar and Cadillac. All of them still had the factory wax on and delivery stickers in the windscreens. The dealers were selling them so fast that there was no time to prepare them properly which meant they were being abandoned by the side of the road because they had run out of petrol.

'It appeared that nobody had explained to the proud new owners that they had to put petrol in the cars to make them go. When the vehicle ran out of fuel the enraged owner would literally leap out and beat it with a stick – just as he was accustomed to do with his camels!'

Everything that was needed for the rapidly developing industries in Qatar was being sent overland at an express rate. Journey times from the UK were averaging fifteen days with some loads getting through in just ten. At this time the sea routes had massive delays of anything up to eighty days because the ports were not properly developed or big enough to cope. It was the same on the west coast of Saudi Arabia where the port of Jeddah was completely clogged up.

To relieve the severe congestion, consignees were given a deadline by which to collect their cargoes. When the date had passed, Gray McKenzie, the Australian port management company contracted to resolve the problem, ordered its workers simply to dispose of everything.

Using some unclaimed bulldozers they pushed the rest of the cargoes off the dockside, straight into the sea! There were cars, buses, all manner of construction equipment, hundreds of tons of building materials, expensive hospital and school supplies, electrical equipment, gas and oil supplies … the list was endless.

Although this was a drastic measure, it did clear the port areas. Indeed, the disposed cargoes helped to make foundations for new quays which provided another source of revenue for transport companies. Some of them based their own vehicles permanently out in the Gulf states and were engaged on internal work transporting cargoes from the ports to inland destinations where construction work was in progress. Michael Woodman, who had already reconnoitred the Gulf, decided to set up his own depot in Qatar to consolidate Astran's position in the region, as described in Chapter 10.

There is no doubt that for many years Astran's overland services from the UK were a major contributor to the development of the Gulf. Not only Astran trucks, but hundreds of others from the UK and all over Europe, were involved in transporting cargoes overland. As David Miller said, 'It is no exaggeration to say that what we carried kept the Middle East alive and developing in those years.'

Even the vast amount of freight being trucked overland was still not enough to supply demand in the Gulf states. So now the volume of unaccompanied traffic – containers and freight-forwarding loads using ships and ro-ro services – began to grow. The Gulf ports were beginning to shape up; some were able to accept more vessels and were beginning to unload them more quickly than before.

In line with this and to strengthen the Astran brand further, Uwe Ploog was recruited in June 1977 as commercial manager. In the *International Freighting Weekly* issue of 1 June 1977 Ploog said, 'My job is to strengthen the company's overall forwarding capability.' Ploog had previously been employed by MAT Middle East where he was overland operations manager.

At the same time as Ploog, Bob 'Haffers' Haffenden joined Astran, starting at the Chislehurst office for a time before moving to Addington. Haffers was experienced in dealing with imports as he'd already been working in the shipping and forwarding business. The majority of his work was dealing with Persian and Afghan carpets which were still making up the lion's share of the return loads. He explained:

'We were still bringing valuable carpets back to the UK, especially from Afghanistan. They had all come via Switzerland because they had to be washed and cleaned there. Some were antiques and others were new, but they all had to be treated at a specialist facility before going any further. Then the loads would be delivered to either Hamburg or London.'

Although there were a lot of personal effects returning from Doha there was not much other work coming back from the southern Gulf states or Turkey, so the Astran overland trucks would travel to Austria or Germany to pick up loads destined for the UK.

Haffers said that he was very impressed with the Astran set-up, especially that the company had representatives across the Middle East: 'There was a local resident in Turkey who would basically say hello and goodbye to the drivers as they went through. There was also a resident in Kuwait, someone in Doha and a resident in Tehran.'

Joining Haffers in the admin department was Alison Moss who was living just around the corner from the Addington depot:

'I worked with Uwe up in his little office for a time doing all his typing and invoices. I'd previously done telex work and so when the telex girl left, I took over her job and also moved into the "cow shed", which was more like a lean-to type of hut fixed to the side of one of the warehouses up in the yard.

'Next to leave Astran was the telephonist so I moved into her office, taking the telex machine with me. I ended up doing all three jobs at the same time. After the bungalow had been extended I was moved into the new reception area and began looking after that too. Then if that wasn't enough, I was given the driver's accounts to look after. I was so keen in those days I even managed to get the job of helping with some of the imports.

'We had some really good times. When some of the drivers came back from a trip they used to bring drink with them off the ferries. They'd sneak across to my office and we'd get bladdered together. Sometimes we'd be there until 10 pm doing paperwork, so it was a great excuse to have a drink in the evenings.

'I remember how things used to grind to a halt if there was a problem. We had a particular phrase we all knew. It was "J F L G P". Whenever anyone uttered those letters, we all knew exactly what to do. They stood for … "Job's Fucked Let's Get Pissed!"'

Not only did Alison Moss work in the office, she was also lucky enough to accompany one or two of the drivers on their trips to the Middle East:

'I thought it would be a good idea so I could get some first-hand knowledge of how it was on the road for the drivers. I went to Baghdad with Jerry Whelan. Trevor Long had arrived there just before us and was ready to come back. By that time Jerry and I had got fed up with each other, so I jumped trucks and got as far as Istanbul with Trevor who looked after me tremendously well. He was a real gentleman. Then I jumped trucks again and came all the way home with John Abraham.

'I thought it was brilliant. It was like a holiday but I also learnt so much about the different cultures and of course the most important thing – the job the men were actually doing.'

Tony Soameson was working flat out to keep the order books full. *Freight News Weekly* 1 July 1977 again:

'There is still a strong demand and as long as overland remains a fast and economical method of transportation, we will continue to maintain our back-up fleet. To inland

Alison Moss accompanied some of the Astran drivers to see for herself how the job went … by all accounts, she had a good time! (AM)

destinations such as Riyadh and Tehran, overland is still less expensive than ro-ro and container – despite increases in transit taxes – and direct road transport remains the method for groupage services where flexibility is essential.

'In most cases inland-bound groupage cargoes for Saudi Arabia are delivered, customs cleared direct to the consignee, by the same driver who loaded in the UK.'

One of the biggest contracts that Astran secured in 1977 was with Ruston Gas Turbines to deliver turbines worth £3 million to oilfields in Iran. A mix of pumping sets and ancillary equipment was packed and loaded onto flat-bed trailers at Ruston's factory in Lincoln and then transported overland directly to Pali-baba and Sabsab in the heart of Iran.

Part of Ruston Gas Turbines' £3 million order is transhipped at Addington onto an extendable trailer ready for the direct run to Iran. (ASTC)

The 3,500 mile journey took on average seventeen days which included a two to three day wait in Tehran while the goods were cleared by customs. Ruston's had used Astran before for various destinations and had no hesitation in using the company again.

Soameson was proud to be awarded the huge contract which saw twenty-seven trailer loads delivered on time. If the same loads had gone via a ro-ro shipping service the transit time would have been much longer due to transhipments and port delays.

During the latter part of the 1970s, Astran fell foul of the transit permits quota system. Some countries limited the number of through freight movements and had their departments of transport allocate a fixed number of permits. Of course there was a lot of hard bargaining on the UK's behalf every year in order to get an increase in the quota.

Astran's problem was that they used up their annual allocations before the year was out. To find a solution, Michael Woodman brought in an expert who had good experience in the shipping and forwarding business. He in turn knew someone who worked for a shipping company who would take the loaded trailers on one of his ferries to the port of Iskenderun in south-eastern Turkey. The trailers would be off-loaded at the port and local Turkish trucks would be sub-contracted to deliver the trailers onward to the Gulf.

Astran paid approximately £200,000 for the deal but then found themselves confronted with a huge problem. Once the Turkish trucks had been booked to move the trailers from the ferry terminal, the truckers refused to drive off the dockside until about £80,000 had been paid directly to them.

It transpired that a forwarding agent for whom the Turkish truckers were previously working owed them all a great deal of money. They were worried that the same would happen again with the Astran loads and were not prepared to take the risk.

The Astran directors and accountants met to discuss the problem knowing that if they paid out all the money their business might well be bankrupt. Bob Paul made a decision to call in the bank which he hoped would be able to resolve the financial problem. 'In fact,' he said, 'we had no choice but to pay the Turks in order to get our trailers delivered - but we desperately needed help from the financial experts.'

A representative of the Hong Kong and Shanghai Bank was dispatched to Addington to discuss the problem. The bank could see that Astran was a reputable company and decided to go ahead and lend the money. Nevertheless, even after the Turks were paid and the trailers had been delivered, the debt was too much for Astran to handle and it was that which started the decline of Astran International.

To add fuel to the fire, as the construction and oil boom in the Gulf states was now in full swing, there were many companies and hundreds of owner drivers engaged on the overland run. Some of these were legitimate, but many were running on a shoe string using illegal documents and poorly maintained equipment. Worst of all, they had no experience of the job but were still keen to have a go because they could see a huge pot of gold at the end of the rainbow.

Exporters were becoming increasingly concerned that their cargoes would not get through to destinations. They had good cause to be worried as many trailers bound for the Middle East were now being found abandoned in places as far apart as Carlisle, London, Rotterdam, Antwerp and even Istanbul. Freight forwarders had given the hauliers thousands of pounds of running money for road tolls and ferries etc., up front, but many of the unscrupulous operators had run off with it never to be heard of again. The trailers and missing loads then had to be traced by insurance companies.

Astran's market was damaged when exporters began to distrust the overland transport system and reverted to conventional shipping container services. On top of this, the Middle East Gulf ports had finally been expanded and were now big and modern enough to handle the huge container-carrying vessels properly.

Throughout this time, the Hong Kong and Shanghai Bank was pressing for its loan to be repaid. However, it became apparent that it was not going to happen. After some very difficult meetings and, very reluctantly, Bob Paul had no choice but to call in the receivers to take control before Astran International went completely bankrupt. As Paul said: 'It was a very sad and sombre day in 1979 when the receivers moved into our offices at Addington.'

GUL 588N and sister GUL 599N loaded with JCB excavators, en route to Doha in the mid-1970s. (JF)

The Long Haul Pioneers

KJN 671P was originally one of three road trains. Note the sand shovel, an essential desert tool, hooked into the ladder. (RG)

CHAPTER 7

The Determination to Succeed

With Astran now in the hands of the receivers, there were immediate redundancies. Peter Cannon said that he was made redundant on Friday but started back with the receiver on Monday. Bob Paul and Uwe Ploog were interviewed and retained as executive directors to run the forwarding and transport operations respectively and some of the general office staff were also kept on as there was still a large amount of work to deal with.

It was at this time that Michael Woodman left the company.

As a very good friend and business partner of Bob Paul, Uwe Ploog put his faith in the troubled business and backed it one hundred per cent. As Paul said:

'Poor devil. Uwe had only been with us a couple of years before the troubles started. He was much more of a businessman than me and he would comment on how relaxed we all were and how the drivers respected me. He'd not seen anything like that before. He told me he couldn't go wrong in backing me and would fight it out with me as he knew the business would survive. That made me a lot happier.'

The trading name of Astran International Ltd couldn't be used when the receiver continued trading. He needed a fresh limited company without any previous liabilities and decided on Astran Freight Terminal (AFT) Ltd, a name that Astran had registered but never traded. Neither Bob Paul nor Uwe Ploog liked the name as it didn't represent the type of business they were in, but neither had any say in the matter.

Ploog wanted to trade primarily as Astran Cargo Services which was already a trading division of AFT used for outside and client relations. After numerous boardroom meetings with the receiver and solicitors, he managed to have the company renamed with his preference. Another name already registered was Astran Truck Services, another division of AFT used for Bob Paul's dealings with sub-contractors and suppliers.

Surprisingly after such an upheaval, Astran continued to operate as normal. Jeff 'Lightning' Woods was taken on to drive the UK shunter because its previous driver, Tommy Thomas, had decided to leave as soon as he heard the company was in trouble. Woods recalls quite clearly the day he started:

'Dave Poulton was showing me the ropes. We'd been to load a trailer and when we got back to the yard we were told to leave the truck outside and go straight into the office where we found everyone else. There was a man standing at the front speaking and he was the receiver. Funny, really, the day I started, the receivers came in.'

Under the circumstances, Woods was a little unsure of his future when he joined Astran. However, he soon settled into his new job solely on UK work which he enjoyed for almost ten years. His little Volvo F86 struggled to tow the heavy trailers, but as Woods said:

Uwe Ploog was able to trade as Astran Cargo Services. (ASTC/CS)

The Long Haul Pioneers

'It was a good workhorse and it always got me there in the end. Before Astran I'd been driving an old Foden truck which had no power steering, no power assistance with the clutch or the brakes and a horrible crash gearbox. When I got into the Volvo it had power steering and a synchromesh gearbox. I thought it was the best motor ever. It was like a Rolls Royce to me.'

Woods's work was relatively straightforward but occasionally he faced problems:

'The trailer had been dropped in the yard ready for me to deliver to JBJ Chairs, a regular customer in London. Over the weekend it had rained heavily and all the water had settled on the roof, making the canvas cover sag down onto the chairs. They had been stacked upside down on top of one another and the legs had then pushed up and through the canvas.

'To make things worse, it was a very cold winter and the whole lot had frozen solid to the roof and to each other. It was a nightmare getting the chairs out and every time we pulled them free we got covered in thick chunks of ice. It didn't do the lovely chairs any good either, as bits kept breaking off. It ruined the beautiful finish.'

Woods knew his job was very important to the whole Astran operation:

'I was always conscientious when I loaded the trailers because I knew they were going a long way and I wanted to make sure the load was properly restrained. Sometimes the trailers were very heavy and more often than not, I'd be right on the weight limit.

'Once I'd loaded, I used to go back to the yard and then fill the big trailer belly tanks with red diesel which in itself weighed four or five tons, so the gross weight of the whole rig suddenly went quite a bit over the 32 ton limit.

'I had a few unusual loads, including some army uniforms in Brighton. The strange thing was that half the uniforms were for the Iraqi army and the other half was for the Iranians. Even stranger was the fact that they were going to be delivered on the same trailer during the Iraq/Iran war. I never knew whether the overland driver delivered to both camps or not.'

The same factory also made the uniforms for the Iraqi police and many of those loads were delivered in one

Jeff Woods drove the Volvo F86 shunter and thought it was a Rolls Royce compared to other trucks he had driven. (SL)

particular Astran trailer, T11B. The Iraqi importer insisted on this trailer being used because it was a metal-box type and supposedly stronger and safer than the normal canvas tilt trailers. It's interesting to note that T11B, designed by Michael Woodman, was manufactured without a chassis due to the low stepped bodywork incorporated for maximum load volume. Therefore it was actually not as strong as the other standard trailers.

On one trip home from the Gulf loaded with personal effects, Barrie Barnes recalled to me that T11B literally snapped in half due to the rough terrain. Barnes had to spend a long period of time sewing it back together with all manner of bits and pieces including wire, rope and steel plates which he had to get welded on.

Although Woods would have liked to have had the experience of driving a trip abroad he was never asked: 'I don't think the bosses wanted me to be away for so long,' he said.

As well as driving his truck, Woods used to enjoy driving a fast car to London to collect visas and other documents for the overland drivers. When things were quiet at Astran, Woods and his little shunter were hired out and could be seen pulling all manner of trailers for regular subbie Mike Taylor Haulage.

164

The Determination to Succeed

Once the receivers had got the company back on track financially, a suitable buyer was sought to take on the business. During the next eighteen months many interested parties made approaches but nothing came of it. In due course a Lebanese company, Gargour & Fils became interested. An agreement was finally made jointly between them and Dubai merchant bank, Wardley Middle East, for the takeover of Astran.

The family-owned company of Gargour & Fils already had the expertise to run Astran because they had similar interests in shipping and transport businesses in the Lebanon. They were aware that Astran was the most respected of all the companies providing overland transport and freight forwarding to the Middle East and they wanted the reputation and specialist service to continue. Work was continually coming in and the financial situation was beginning to improve again. More employees were taken on and new trucks were ordered.

Since the first Scania Vabis trucks, UHM 25F & UHM 26F, had gone into service with Asian Transport in 1968, the Swedish manufacturer always had a special relationship with the company. The fleet was totally reliant on the Swedish marque apart from the Volvos hired from Avis.

Of course sub-contractors had their own preferences and were driving all manner of trucks, but those who chose Scanias had the added advantage of being able to use Tony Keirle's expert services in the Astran workshops at Addington.

To try and infiltrate the Scania fleet, truck salesmen from other manufacturers were constantly knocking on Bob Paul's door with eye-catching deals, but he was adamant that he would never use anything but Scania. Nevertheless, Paul eventually took the decision to try Mercedes Benz after a salesman offered him a deal he really couldn't refuse.

The result was an order in 1980 for three brand-new Mercedes tractor units, badged as 1632, each costing about £27,000. They were delivered as left-hand drive versions with double-bunk sleeper cabs, air-conditioning units, extra-large fuel tanks and special front-mounted sand filters for dusty conditions.

The purchase did not go down too well with the drivers, as everyone – and especially chief mechanic Tony Keirle – was a committed Scania man. They also knew of the bad reputation of the German trucks, renowned for being sluggish and uncomfortable.

As executive transport director, it was Bob Paul who had

One of only three Mercedes 1632s in the Astran fleet, NKM 284W, driven by John Lewis, is resting in Saudi. (JL)

MKN 681W discharging cargo at Addington after a gruelling journey home from the Middle East. (ASTC)

the final word. He decided to sell three of the older Scanias to owner drivers and replace them with the new Mercedes trucks. They were handed to Jimmy Hill, Nick Carter and Mick Henley.

As 1980 came to an end, business was going well for Astran, impressing the new owners. Relationships were excellent and Paul remembers fondly referring to the Gargour & Fils directors as 'the Gar-gars'.

Paul and Ploog were each rewarded with 5 per cent of the company shareholding while Gargour & Fils held the remaining 90 per cent. Ploog commented that he saw a bright new future for the company:

'An immediate priority is to consolidate our direct overland services. We have already taken delivery of three new vehicles and expect to make a capital investment in the region of £300,000 during 1981. With the backing of our financially very sound shareholders and with renewed confidence already evident amongst customers, we expect to show substantial gains in business over the coming months.'

As Ploog expected, business did continue to increase.

In 1981, when Paul returned to his friends at Scania to replace the Volvo F86 UK shunter, he was offered a demonstrator for the knock-down price of around £4,000. Like its Volvo predecessor, the Scania was small and lightweight.

Instead of a sleeping compartment it had a bunk which had to be folded out at night once the seats were folded down. Although Woods was disappointed that his new truck didn't have a proper sleeping arrangement, he soon got used to it. The original shunter was demoted to moving trailers around the depot.

Along with the new shunter, three bigger Scania 142s – AOO 66, 67 and 68X – were delivered for the overland fleet. They all had the latest V8-powered engines and new-shaped cabs. Previously, the majority of trucks in the fleet had smaller 11 litre engines, but Paul justified having the bigger 14 litre V8 engines because the extra power and torque would cope more easily with the arduous terrain in Turkey.

One of the new trucks, AOO 66X, was given to Dave Poulton who was especially pleased with it. He found the truck to be exceptionally reliable and clocked up over 650,000 kilometres in it before he was redundant in 1988. The truck was then sold to owner driver Mike Walker who fitted it with an extra axle and then clocked up another 300-400,000 kilometres!

The truck replacement programme continued. In 1982, John Bruce arrived back at the depot in his beloved Scania UAR 747S to be met by Bob Paul who announced that he and Rick Ellis were about to get a new truck each. Bruce takes up the story:

YTW 875X new at Addington. Bob Paul paid about £4,000 for the little Scania. (SL)

'One was a V8 Scania and the other was a Mercedes. He said he was going to toss a coin to see who had which one. To be honest, I was disappointed because I knew about the other Mercs and didn't really want one after my Scania. Typically, I lost the toss and got the Merc while Ellis, the lucky sod, got a new Scania, EKO 950Y.

'Bob and Peter Cannon took me to Sparshatts, the Mercedes dealer in Sittingbourne, but as soon as I walked into the garage I got a pleasant surprise. Standing before me was a brand-new shiny Mercedes tractor unit, GKJ 82Y. It had the latest cab and a bigger engine than the ones we already had. It was one of the very first in the UK and I was over the moon.'

Bruce presumed that the other drivers' complaints about their Mercs had been noted. His new Mercedes cab was much larger than the others and the engine was the latest V8 version which was far from sluggish.

When Bruce took his new steed on its first run to the Middle East he met Dick Snow driving a Scania at Istanbul:

Dave Poulton clocked up 650,000 kilometres in AOO 66X. She is taking a well-earned rest on the coast at Doha. (DP)

'Snowy was the first one to see my new Mercedes and I remember feeling quite proud as I pulled up alongside him. He didn't really say a lot about it, but when we went up the Bolu pass between Istanbul and Ankara, he just couldn't keep up with me. His truck was also a V8 but even though it had more horsepower, it didn't have the torque that mine had.

'When we got to Ankara for a rest stop, he suddenly showed

Problems such as those caused by the Iran-Iraq war seem far away from this part of the fleet posing for publicity photographs at West Malling airfield, Kent in the early 1980s. Left to right: WHJ 194S, WHJ 193S, NKM 300W, AOO 67X, MKN 681W, KJN 671P and SOO 815R. (ASTC)

Sub-contractor Mike Walker bought AOO 66X from Astran in 1988 and added the extra trailing axle. (UNK)

The brand-new AOO 68X posing for publicity photographs at Dover. (ASTC)

The Determination to Succeed

John Bruce, driving Mercedes GKJ 82Y, ran with Dick Snow who couldn't keep up in Scania AOO 68X. (JB)

interest and couldn't resist a good look round the cab. I nearly fainted when he said he wanted to swap his Scania for my Merc. I was chuffed to bits and soon after I became a converted Mercedes fan. In fact I became envied by most of the other Astran drivers who all wanted to swap with me.'

A second new Mercedes, GKK 906Y, was handed to Mike Walker.

Not wanting to be outdone by the German manufacturer, Scania managed to place three more 142 V8 tractor units, EKO 948, 949, 950Y, into the Astran fleet at that time. The two Mercs and the three Scanias were the final vehicles that Bob Paul bought while he ran the own-account transport fleet.

By now, Astran was concentrating heavily on the southern Gulf states. After the Shah was deposed, business to Iran and Afghanistan had stopped immediately but trade to Qatar took off and was soon followed by an upturn in trade to Iraq. Even with Saddam Hussein in power, trucks were making regular deliveries to Baghdad and the country also became the main transit route to the Gulf.

Bob Haffenden recalls a very lucrative contract with Hyundai in which Astran delivered two hundred loads of road markings deep into Iraq for a new road which was being built between Jordan and Baghdad. Each trailer load was delivered direct to the section of road being constructed and deemed urgent with a capital U.

Ten or twelve loads were leaving the UK every week and were mixed between UK drivers, Bulgarians and Poles. Joe Bowser, one of Astran's drivers, still has clear memories of the job:

'My trailer was loaded with 45-gallon barrels containing a tar-like substance. The barrels were bouncing around as I drove along the rough roads, and the bottoms began to pop out. As my trailer was customs sealed, there was nothing I could do about it.

'Every time I stopped, I noticed that more and more tar was leaking through the trailer floor and sticking to the truck. It was all over the wheels, lights and chassis. It was everywhere and as hard as I tried, it was impossible to remove it. The weather was hot which helped a little, so I used a scraper to try and clean it off but just gave up after a few minutes as I was getting nowhere.'

Bowser was heading to one of the construction camps south of Baghdad and could not get there quickly enough. He said it was all a bit hit and miss with vague delivery details:

Joe Bowser, driving GUL 588N, finally found 'something' miles from anywhere in the Iraqi desert. (JBO)

'The telex instruction was very brief: "Go to Baghdad. Clear customs. Head south to Basra. When road ends turn left. Head out into desert and you will find something."

'After I'd cleared customs I was able to release the trailer seal and examine the load, but it was far too late to do anything because the tar had completely emptied from some of the barrels so I just followed the telex instructions.

'Looking for "something" in the desert was like looking for a needle in a haystack so I just kept driving ... and driving ... and driving until eventually I spotted a few workers and one huge excavator. That must be it, I thought, and as luck would have it I was in the right spot. They used the machine to lift the barrels out of the trailer and what a mess it was when they had finished!

'As far as I remember, the tar was still all over the truck when I was assigned to a newer one a good few months later.'

Tony Soameson was by now a full-time subbie running his own truck in Astran livery. His favourite destination was anywhere in the Gulf. 'I loved it down there,' he said. 'The people were fantastic. I'd done Iran and Afghanistan in the early days but the Gulf destinations were so much better in my opinion.'

Unfortunately for Soameson, the authorities didn't always agree that he was so fantastic. On one occasion he volunteered to take a very large Portakabin from Abu Dhabi to Aqaba in the south of Jordan, a journey of about a thousand miles. The cabin was wide and long, so Soameson rigged up some extra marker lamps on each side to warn other motorists of the width.

With his orange roof beacons flashing and all his

No matter how hard Bowser tried, he couldn't remove the sticky tar from his truck! (JBO)

Tony Soameson waited for darkness to fall before heading off with the extra-wide load. (TS)

other lights full on, he drove through the night when there was less traffic. He had almost reached the end of the TAP Line in Saudi Arabia when he was pulled over by the police who made it very clear that he should not have been driving such a wide load at night.

Soameson was outraged. He knew that the locals carried similar loads pitch black, without any lights at all, and it would not be possible to see that they had wide loads until it was too late and a serious accident had occurred. The police were not interested in Soameson's plea and escorted him to the nearest police station where they made him stay until daylight.

Luckily he was not fined and actually made the most of his unforeseen break by chatting to the policemen. He got on so well with them that he ended up drinking chai tea with the police chief in his luxurious office.

By the mid 1980s trailers were leaving the UK almost daily on an express service to the Gulf. Of this traffic, 35 per cent was heading to Qatar, 25 per cent went to Saudi Arabia, 20 per cent to Iraq and the remainder was split between the United Arab Emirates, Jordan and Kuwait. The state of Qatar became a very lucrative market and new contracts with companies working on the gas and oil refineries meant that Astran was still in constant demand to deliver cargoes as quickly as possible.

There were also some isolated destinations to which Astran sent trailers. On one ultra-long trip, Frank Hook, in Mercedes NKM 284W, was running in convoy with Geoff Frost who was driving one of the oil company's own rigs. The pair collected oilfield

Vision behind was completely obscured by the extra-wide cabin! (TS)

equipment in Germany and drove it to Muscat in Oman, a round trip – including delays – of eight or nine weeks.

Unfortunately, there was an error with the German documentation which meant the men were not allowed to unload in Muscat. 'Actually,' said Hook, 'it was fantastic':

SOO 815R about to depart the UK for Qatar loaded with urgent accommodation for construction workers. (ASTC)

'At first we had to stay in the trucks, but the British ex-pats who were working there took us out to the restaurants and clubs. Once we knew we'd be stuck there a while, we were invited to take rooms at the hotel. How could we refuse?

'Geoff and I had our own rooms and when we went down for dinner all the waiters were fitted out in immaculate uniforms. It was silver service too. It makes me laugh now to think about two English drivers wearing mostly jeans or shorts and t-shirts staying for ten days at the most expensive hotel in Muscat. We got some funny looks from the other residents who obviously had loads of money.'

Saudi Arabia's construction and oil industries were experiencing a huge amount of development which was very reliant on overland trailers to feed its needs. Most of the machinery cargoes were not properly suited to being shipped in sea containers. There were many awkward-sized and delicate loads which needed a direct road delivery to each site to save time and reduce damage.

As there was now an unprecedented amount of work coming to Astran; the overland drivers were working flat out, aiming to arrive in Riyadh or Dubai within 12-14 days after leaving the UK. It is fair to say, though, that most journeys were done in less time.

Peter Cannon said the he now found himself dealing not only with the Astran trucks but also with the ever-

NVW 488P pictured at Addington loading machinery spares for the oil industry in Qatar. (TS)

growing number of sub-contractors hired in to cope with the increasing workload:

'There's nothing worse than asking a busy man to do more. The mid-1980s was our busiest time and we had something like seventy-eight trucks on our books. I think we had fifty-six of them outside the UK in one week alone. I had an awful amount of work to deal with.

'We had to treat most of the subbies as our own drivers. It just wasn't possible to ask them to arrange their own visas or TIR carnets. It wouldn't have happened. It was bad enough trying to get invoices from them because all they wanted to do was drive their trucks up and down to the Middle East.'

Gargour & Fils seemed very happy with the way the business was coping and had always taken a back seat, allowing Paul, Ploog and their team to run the company without interference. Then out of the blue at one of the regular board meetings it was suggested that Paul stopped renting the premises at Addington as he was seen to be wasting valuable money.

Gargour & Fils announced that they would buy the depot from the landlord as they saw it to be a viable asset. After much negotiating, the Middle East freight terminal was bought for a little under £300,000.

Over the next couple of years, the forwarding side of Astran's business continued to do very well as it was easy to make money that way. However, the transport operation began losing money mainly due to the knock-on effects of the political and religious climates in the Middle East.

On the long haul to Muscat, Frank Hook drove NKM 284W while Geoff Frost (pictured) took the oil company's own vehicle, seen in the rear. (FH)

Dave Poulton delivered this 32 ton electrical transformer to Damman in Saudi Arabia. They are ready for the off at Addington. (DP)

Whereas shipping container services had by now become cheaper and much more reliable, overland transport was becoming fraught with problems. For example, the Iraqis, Syrians and Jordanians were starting to put their transit taxes up by vast amounts. These had to be paid before any rig was allowed to drive through their countries. They were worked out according to the value of each load and it was not unusual to pay £300-£400 for a single day's driving. It

Unloading at Damman. (DP)

soon became very expensive to send cargoes overland.

To makes things worse, the Iraq–Iran war brought the closure of Iraq's main ports so that the country had to rely solely on overland transport. Many UK owner drivers now had a go at the Middle East run because of a serious lack of domestic work.

Desperate to poach loads, some unscrupulous UK operators were taking on work to Baghdad for £3,000 or less, compared with Astran's rate of around £5,500. Astran didn't believe that anyone could make money on the slashed rates and they found their work drying up just as it had done some years previously.

An inflating price which contributed to Astran's problems was the cost of repair and maintenance for the trucks when they arrived home. This had increased to about £700 per truck per trip.

As Peter Cannon was struggling to find loads to the Middle East which paid decent money, he began to make enquiries about other suitable work so that at least the vehicles would be kept earning. 'We didn't know what was going to happen with the overland routes, so it was a case of spreading our wings and doing something else to earn money,' Cannon said.

Scania 140 V8, WLE 344M, parked at Addington loaded with machinery destined for the Gulf. (PC)

Ferryline Trailers at Dover agreed to take on one of the trucks as a contract vehicle. John Bruce was told that Astran would still pay him, but he would be towing Ferryline trailers for a time. Bruce said that he enjoyed the 'local' regular trips between Bridgend in South Wales and various Ford car plants in Germany. The arrangement with Ferryline lasted for a year or so.

By 1988 Gargour & Fils had grown into a colossal organisation and had lost interest in Astran, preferring to concentrate on their huge shipping empire. Notice was then given that they were going to put the whole company and the Middle East freight terminal at Addington up for sale. Bob Paul said that their decision to do that was based solely on the loss-making transport operation. Wasting no time, Gargour & Fils washed their hands of the business and the depot which was sold to a property development company at closed auction for about £800,000.

Paul had already warned his drivers that they should be very careful of the escalating running costs and they should try and be more economical in all areas because the owners wanted to close everything down. He fought tooth and nail to save the company which he had been part of for over twenty years.

However, after numerous meetings with the owners, he had no choice but to close the transport operations as it was in so much trouble. 'They were all accountants,' Paul said. 'They weren't interested in the transport, only the money. As we were just loss-making, we stood no chance.'

Paul and Ploog wanted desperately to keep the whole company running. They negotiated a deal by which Gargour & Fils kept the money from the sale while they would keep the majority of the shares and be in total control again.

As part of the deal, Paul and Ploog were required to put up a large sum of money. Paul was not really interested in owning the company and agreed that Ploog would pay the whole sum, making him the new owner of Astran.

The business would continue to concentrate on freight forwarding but would have no employed drivers or trucks of its own; instead it would rely solely on sub-contractors. All of Astran's employed drivers were therefore made redundant.

This caused immense distress to Bob Paul: 'I am not a suicidal person,' he said, 'but if I were, I would have killed myself when I had to shut everything down. I felt terrible. It broke my heart.'

Dave Poulton telephoned the office on his way home through Turkey. A very sombre Peter Cannon told him that when he got back he wouldn't have a job unless he wanted to buy his truck and run it as an owner driver.

This didn't interest Poulton so he finished working for Astran on his return. He started with another company almost immediately, driving a much smaller truck. 'It was right-hand drive,' he said. 'I spent the next six months driving down the pavements because I'd spent nearly seventeen years in left-hand drive trucks.'

Once the premises at Addington had been dealt with, it was a frantic hunt to find a suitable place to continue trading with the freight forwarding and shipping business. As luck would have it, regular sub-contractor Mike Taylor had a place not far away at the Platt Industrial Estate in Borough Green. Taylor rented his warehouse to Ploog and agreed to handle all the warehousing. On 20 June 1988 Astran Cargo Services upped sticks and moved out of the Middle East Freight Terminal, relocating straight into Unit 9, Platt.

Taylor sited a large Portakabin outside the warehouse for use as an office. The huge downsizing move from Addington was successful enough for Ploog and Paul to run the business more or less uninterrupted. Jan Payne remembers her new office well: 'We were in a bit of a dip and every time it rained we ended up with a moat all around us.'

Unfortunately, the landlord took exception to the Portakabin as no one had asked his permission to put it there. He made it clear that it should be removed immediately – so without further ado, everyone had to move yet again and squash into a much smaller office which was built inside a corner of Taylor's warehouse.

While Uwe Ploog put all his efforts into the forwarding side of the business, Bob Paul wanted desperately to keep his interest in the transport operation alive. Now that his immediate family of employed drivers had dispersed, he wanted to concentrate on the extended family of subbies. As Paul said: 'They might have been owner drivers, but they were still my boys.'

After Paul's employed drivers had been given the grim news of redundancy, they were asked if they would consider becoming owner drivers. Paul explained:

'We still had loads to get to the Gulf and I wanted to use drivers who I knew and trusted. I put it to some of them, including Mike Walker and Rick Ellis, that they should become owner drivers and continue to work for me.

'I held the company documents, permits and so on, so I suggested that they should buy a truck and I would finance them. I would supply permits and transfer everything properly into their names, even getting operator licences sorted out so they would not be beholden to Astran at all.

'The deal was a truck and trailer, both of which were fully serviced, and it also included permits, etc. The trucks were worth about £10,000 each, and they were to give me £500 from every trip to pay me back. I saw it as a way that we could all make money and I felt we could get back properly into overland transport again.'

The Long Haul Pioneers

Smarty pants SOO 816R brand new at Scantruck in Essex awaiting delivery to Astran. (PS)

WHJ 194S and trailer both looking worn out at Addington. This was in the transition period of changing back to red-and-yellow livery from ochre and white. (SL)

Tony Soameson, standing right, discusses the menu in a favourite restaurant near Tarsus, southern Turkey. Note the fresh, hanging meat. (TS)

By the end of the 1989, Astran Cargo Services had re-developed its business. The forwarding division was very successful and, surprisingly, the overland transport division was also holding its own despite the political and religious problems confronting them. 'We were doing very well again by then,' said Paul:

'The work seemed to be pouring in and the owner drivers had settled into running their own businesses, keeping the trucks very busy. However, we couldn't work in the ridiculous conditions of the tiny warehouse and office. I knew that another unit on the same industrial estate was available so I said to Uwe, "We have to take a chance on this because we've done okay up until now. We've done a few trips to the Middle East and made some money."

'I told him I was no good with finance and that he was the expert, so he should be the one to talk to the bank manager. The price for Unit 6 in the Platts Industrial Estate was about £400,000. Amazingly, Barclays couldn't have reacted more quickly. This allowed us to move more or less straight into the bigger and better unit. We put up £100,000 and they gave us a mortgage for £300,000.'

Paul told me that the money lent by the bank was paid back within seven years:

'Neither Uwe nor I was interested in taking a huge salary. We were perfectly happy working the way we were and all we wanted was enough to live on. We kept the staff salaries good and managed to make a go of it at the new depot.

'I was still totally committed to the business and had the determination to succeed with it, especially after everything that had happened over the years. This was my life and I was not prepared to let it all go overnight.'

CHAPTER 8

Business as Usual

With Astran relocated to Unit 6, Platts Industrial Estate, Bob Paul felt relieved that he and Uwe Ploog were able to continue trading in the business they loved so much. They had come through the financial problems and were starting to pick themselves up again. The business was doing well and, best of all, they had the full backing of their office staff and loyal support from the sub-contractors.

Now that there were no more own-account drivers it did seem awfully quiet. The one thing Paul missed was the banter with his drivers and seeing them coming and going from the depot ... and the pub!

Because they relied solely on subbies for overland transport, most of the full-load cargoes were being collected away from Platts, the trucks then going straight to the docks. When the subbies returned to the UK, they would deliver their load and then return to their base without going to Astran's HQ. Obviously there were occasions when they had to visit Platts to pick up smaller loads, groupage, or to sort out finances, legal documents and so on.

Some drivers just called in to see everyone. As Paul said, 'I really looked forward to that, it was like old times. Getting into the new routine of operating with only subbies certainly took some getting used to.'

As the early 1990s progressed, it became clear that there were political tensions in Europe and war in the Middle East was looming. However, Bob Paul had seen this kind of trouble before:

'I was not fazed. There had always been problems and wars somewhere along the line in all the years that I had run my trucks, so this was no different. We just carried on as we always had, running the business in a normal fashion. We had loads to deliver and customers waiting.'

Terry Tott was unfortunately caught up in the start of the Iran-Iraq war in 1980:

'I arrived in Baghdad on 16 September 1980 with a consignment for the International Trade Fair. I spent a few days clearing customs and unloading, and on September 22nd I heard about the war on the long-wave radio. Along with some other, non-Astran, drivers I went to the British Embassy who told us not to worry. We went back to the trucks and had a drink and a laugh about it.

Tott was woken abruptly the next morning by the sound of rapid machine-gun fire:

'I looked out of my window and saw streams of tracer bullets going across the sky. If that wasn't bad enough, a Phantom jet shot past and fired a couple of missiles. To be quite honest, I was petrified.'

Tott and his colleagues tried again to get help from the British Embassy but were told to stay in the safety of the Trade Fair, so they spent a second night parked in the compound, fearing for their safety. The next morning Tott was again woken by fighter jets but this time he witnessed one of the jets being shot down very close to his position:

'That was it. I had to get out no matter what. I waited until nightfall and then drove flat-out without any lights for about 150 miles to the border with Turkey.'

Tott had no lights because of the black-out order. En route he picked up a British woman with two children, giving them a lift until they caught up with a convoy of British evacuee buses.

'When I got to the border there was a massive long queue but I was so concerned about getting out of Iraq that I just kept driving past all the other vehicles and people. I just had to get out of there.

'I had a bad time getting the authorities to let me out at the border post, but after a lot of arguing and telling them I was British they finally let me go.'

Gavin Smith, an Astran subbie for about ten years, spent the majority of that time on routes throughout Europe and to the southern Gulf states, so he had plenty of experience of political trouble spots:

'By the time 1990 came along, I was already pretty well seasoned on the Middle East run and knew my way around so I was easily able to divert if there was trouble ahead. My main routes were from either the UK or Europe to Doha, but there were also occasional loads to Deir ez-Zor and Damascus in Syria, Kuwait, Bahrain, United Arab

Emirates and some far-out places in Saudi Arabia. I was constantly kept busy and was actually earning quite a decent living out of it, even during the troubled times.'

John Harper is another very experienced subbie who knew the routes like the back of his hand and how political situations could change dramatically from day to day. He was on his way to Doha at the time of the first Gulf War, but was running late as he was baby-sitting a driver who had not been to Doha before. Harper was struggling to motivate his companion and decided to leave him at Habur on the Turkish exit border into Iraq, telling him to make his own way onward.

Seasoned driver on the Middle East run, Gavin Smith (sitting left) enjoys a coffee break with Lebanese friends Sammi el-Diddi and co. (GS)

Harper and another subbie, Mike Walker, entered Iraq on 2 August 1990, the very day that Saddam Hussein's forces invaded Kuwait.

Harper and Walker had already queued for a lengthy period and completed their customs formalities to leave Turkey at Habur. Technically they were in no-man's-land and could not turn back. They had no choice but to continue into Zakho, the entry border into Iraq. After being held by officials for a number of hours while everything was scrutinised and their explanation that 'they were just truck drivers who were trying to do a job' was verified, they were finally let go.

Harper made speed through Iraq without any problems and arrived during the night south of Basra at Safwan on the Iraq/Kuwait border. He cleared the exit border and then for security was escorted in convoy to the fruit market in Kuwait City, the normal customs clearing point for trucks delivering to the country.

Harper was told to park there overnight. His passport and documents were placed inside the customs building for safe-keeping until the convoy resumed the following day. Normally the convoy ran straight through Kuwait to the border with Saudi Arabia at Khafje in one go, but the authorities said it was not possible because of the war, hence the overnight stop at the fruit market.

The next morning, Harper was woken very abruptly at 5 am:

'All hell was let loose. There were helicopters flying over and tanks roaring past. When I looked out of the cab, I could see they were Iraqi forces and I suddenly realised what was happening. Kuwait City was being invaded right in front of my eyes!'

Harper ran to the customs office where his papers were being held. The building had already been hit by Iraqi tanks and was in ruins, but Harper was unperturbed and went inside. Luckily he found his possessions and set off hell-for-leather towards Khafje. When he got there the place was an absolute shambles with hundreds of people trying to escape the country. Harper managed to talk his way through and made it safely into Saudi Arabia.

Meanwhile, Mike Walker had negotiated Baghdad during the night without any problems but chose to divert to the south-west across Iraq to Ar'ar on the Iraq/Saudi border. He was very experienced on the Middle East run and had seen plenty of military action during his time driving at the time of the Iran-Iraq War in the 1980s.

When Walker arrived at Ar'ar in the very early hours of the morning, he was refused an exit passage by the Iraqi forces. He pleaded with them and showed his visa stamps to prove that he was a genuine truck driver.

After a while he spotted an Iraqi official whom he knew from previous trips and with his help was able to convince the authorities that he was only trying to pass through the country and meant no harm. Eventually he was permitted to continue his journey and drive unhindered to Doha where he met up with Harper.

The pair rang their wives who were understandably worried as they hadn't heard from the men. Harper said, 'My wife told me about the invasion of Kuwait and asked if I had been involved. I just said, "No, love, not me."'

Harper also rang Bob Paul who was extremely concerned for all his subbies:

'Bob asked about the driver I was supposed to be escorting. I told him that I'd lost touch with him, but I would come back via a different route through Saudi, Jordan and Syria and look for him along the way.

'When I got to the town of Hofuf in Saudi Arabia which was a good day and a half's drive from Doha, I spotted the man's truck parked on the desert. I was quite concerned to see it with doors and windows open but no sign of the driver.

'I got into the cab and found his passport which had stamps in it to say he had come through the border at Ar'ar three days before. To be honest, I thought he'd had enough, dumped the truck and done a runner.

'The locals told me that some Turks had taken him off for a meal nearby. I was furious and when he returned I made my feelings quite clear to him. I also told him to contact Bob Paul as soon as possible because he was pulling his hair out worrying where he was. The bloke just didn't seem to realise the importance or the danger of everything.'

Harper continued his homeward journey, re-loading in Germany for a delivery in the UK.

'When I got back, I called Bob Paul to ask what was happening and he told me he wasn't sure if there would be any more trips for the time being. He did, however, give me an extra £500 bonus for running through the war and getting through to Doha.'

It was a very stressful time for all the drivers and for the Astran staff back in England as no one really knew what would happen next. Wives were constantly ringing Bob Paul for news, and he in turn was relying on the British Foreign Office for updates:

'They gave me some very good information. They said whatever you do, don't go down there. Then they asked if I was going to pay all the drivers' mortgages and things like that. Another route around the war zone was available, so I told the men to go that way instead. It was longer, but not too bad except it was more expensive due to transit taxes imposed by Syria and Jordan.'

On Saturday 18 August, Mike Walker also made it home. Not put off in the slightest by the escalating war, he telephoned Astran's office the following Monday to enquire when the next load would be ready for him! A week later, Walker was loaded and on his way to the Gulf again.

In an interview in the October 1990 *Truck* magazine Walker explained that he was going back out because as long as he kept driving to Doha he could just about keep up with his mortgage payments: 'Financial responsibilities mean I cannot stop just because a lunatic such as Saddam Hussein wants to chuck his weight about.'

Walker commented that although his ageing truck was paid for, he couldn't afford to replace it. Getting hire purchase on a new one would be financial suicide because the insurance companies considered the whole Gulf region to be a war zone: 'If the truck gets blown up, then it's not covered. Yesterday I tried to increase my life insurance but they practically fell about laughing.'

'Business for Astran was very good during the war,' Paul said. 'We were the only ones doing the overland runs to the Gulf and, with very long delays for ships, the war did us a huge favour.'

Although work for those subbies who wanted it continued as normal, one or two of them refused to take loads, while others made it as far as Ankara and said they couldn't go any further. Some got through Turkey but were refused entry into Syria and others were turned around at the Jordanian border.

Mark Stewart remembers leaving the UK on a trip to Doha just one week after the Gulf War started in August 1990:

'I was assured by Astran that I would be okay so long as I took the officially advised route and stayed well away from Iraq and Kuwait. I was also given £500 bonus which helped the situation.

'When I got to Ankara, I heard that four or five British trucks had managed to get through to Doha so that gave me the confidence to go for it.'

Stewart continued unperturbed and completed his round trip to Doha successfully. He returned there again on more than one or two occasions but always managed to avoid the war zone:

Insurance companies would not touch the likes of Mike Walker if his truck got blown up. Gavin Smith's rig is posing by some discarded missiles! (GS)

'After a time, the Palestinians were getting a bit fiery at Ramtha, the entry border into Jordan, and began stoning the trucks as they went past which was very unpleasant. It was at the time when Scud missiles were being fired at Israel and the people were obviously very angry with the situation because Turkey was allowing the United States Air Force to use its air bases.

'The locals could not distinguish between Western trucks and Turkish ones which were who the stones were really aimed at. We just happened to be there too, but luckily for me there was hardly any damage done to my truck.'

Did any of these war experiences ever put Stewart off the job: 'No,' he said. 'I was young and daft.'

Technically, none of the Astran trucks would have been in the war zone so long as they followed the official advice to drive from Turkey through Syria and Jordan, then via the designated transit route through Saudi Arabia, well away from the Trans-Arabian Pipeline (TAP) which was relatively close to Iraq and Kuwait.

The designated route took them centrally into Saudi to Ha'il then Buraydah and on to Riyadh which put approximately one and a half days onto the journey. John Harper recalls trying a better route via Medina and Mecca but unfortunately was not allowed near either town because he was not Muslim.

'I was stopped by the police as I tried to get round the Mecca ring road,' Harper said. 'They weren't interested in any excuses and just turned me straight round and escorted me well away.'

Like the other subbies, Harper found himself with plenty of work hauling trailers destined for Doha. He avoided Iraq and Kuwait entirely:

'I remember arriving at Haditha on the border of Jordan and Saudi Arabia just as the air war was starting. I had never seen it so busy there as all the local traffic was turned back into Jordan and it left just me and Andrew Wilson Young stranded in the huge border compound. We were kept there for about three days but eventually the officials let us go because they were getting very jumpy and were probably ready to get out of the place themselves.'

Gavin Smith loaded in the UK a few days after New Year's Day 1991 and along with several other subbies – Nick King, Mike Walker, Alfie Foulkes and Bob Poggiani – was en route to Doha.

The Long Haul Pioneers

Mark Stewart completed regular trips to Doha during the first Gulf War. He is in convoy waiting to enter Doha during the conflict. (MS)

Arriving at Istanbul, Smith phoned Bob Paul and was told to head for the Telex Motel truck stop at Ankara and await further instructions. By 14 January the allied bombing campaign for the liberation of Kuwait was about to start. Smith and the others were now parked in the motel where they spent five days waiting for information.

'During the wait, we were hearing all kinds of horror stories about what was happening to trucks and their drivers at Cilvegözü, the Turkish exit border into Syria, and beyond. After five days just sitting and waiting, I decided that I was going to head off and would go as far as possible. The rest of the guys agreed, so we all set off.

Waiting for the Gulf War air campaign to start in January 1991. Left to right: Gavin Smith, Alfie Foulkes, Mike Walker, Nick King, mate of Nick King, Bob Poggiani. (GS)

Our own Astran convoy: Gavin Smith and the others heading south on the detour transit route through Saudi Arabia. (GS)

'Apart from the road being quieter than usual, and loads of war-plane activity around Ceyhan, the US Airbase near Adana, the journey was good. When we got to Cilvegözü, we were the only trucks there except for a couple of Eastern Europeans that were going to Amman in Jordan. All the Turks had parked their trucks in no-man's-land and had walked back into Turkey.

'We managed to get out of Turkey okay and entered the customs area at Bab al-Hawa on the Syrian side - much to the surprise of the guards. We submitted our paperwork as normal and then set off in the safety convoy as usual across Syria. We got to the exit border of Dara on the Syria/Jordan border but then we couldn't go any further. There was some tension there because a lot of the people in Jordan were Palestinians who didn't like what the Americans and allied forces were doing, so we weren't allowed into the country.

'We sat in the convoy compound for almost a week. During that time a convoy did depart to try and cross Jordan, but was turned round after it was stopped by an angry mob who managed to upturn two Turkish fridge trucks!

'Eventually things settled down and we got through the Ramtha entry border and set off in convoy across Jordan to the exit border at Al Umari where the tension was even worse. After keeping a low profile and completing our Jordanian exit formalities, we departed across the stretch of no-man's-land and entered Haditha, the entry border into Saudi Arabia.

'The Saudis were very surprised to see us, but nevertheless our customs formalities were completed and after having our loads thoroughly inspected, we were allowed to enter the kingdom. From then on, we followed the designated transit route and drove unhindered right through to Doha.'

On the return leg to the UK, all drivers were ordered to sign a document at Salwah on the Qatar/Saudi border saying they would use the designated route at all times on the way back. However, some subbies blatantly ignored this and ran the gauntlet up the TAP Line. As Mark Stewart explained:

'We thought it would have been safer if the authorities had allowed us to follow the TAP Line instead of making us use the long detour. There was so much military traffic up and down there that we would have been looked after. I remember coming home from Doha and disregarding the official route, taking a chance on the TAP Line instead. Maybe not a good idea now, but I went for it.

'There were American military camps all along the road, all waiting to go into Kuwait at a moment's notice. I used to have regular visits from their Apache helicopter gunships that were obviously checking me out. Some of them were flying at cab level height and the noise was unbelievable. I'll never forget that.'

Nigel Harness was working for Peterlea Trucking, one of Astran's subbies, at the beginning of the Gulf War but, luckily, he was delayed when the gearbox on his Volvo Globetrotter gave up in Germany. He was originally running with Mike Walker and would have stayed with him through Iraq and onward to Doha:

'Because of my gearbox problems, by the time I got to Turkey I had heard of the invasion and had to divert through Syria, Jordan and Saudi. My boss Peter King would not let me carry on after that so I wasn't one of the ones who ran through the war. But I would have gladly done it myself because that was the challenge of the job – and believe it or not, I really enjoyed that.

'Peter wasn't worried about personal safety. It was the risk of whether all the borders closed and then the truck would have been trapped and he wouldn't have made any money.'

During February 1991, Astran had about eighteen drivers in or near the war zone, but all of them made it to Doha and back to the UK without any problems.

The first Gulf War finished on 28 February 1991 and then during the summer the first loads went back into Kuwait. Drivers reported that customs formalities were very informal compared with those before the war and the border control was virtually non-existent. In fact all it took was a stamp on a passport, a glance at the official documents and most importantly, a large slap on the back from the guards in praise of British help, and the men were free to drive into the country.

Driving at night, the oil fires could clearly be seen burning for many miles, evidence of the terrible damage done to the country. The roads which used to be new and smooth now resembled the surface of the moon with craters and potholes everywhere.

The dual carriageways had been bombed so badly that only one side was passable – and only with the greatest of care – and many burnt-out tanks were left abandoned all over the place. Some of the towns were totally abandoned and although the buildings were not all demolished, it was very clear that they had suffered a good battering.

One of many burnt-out tanks abandoned by the highway. Close by is Nigel Harness's rig contracted to Astran from Peterlee Trucking. (NH)

Nigel Harness's Astran rig is parked with two other Peterlee Trucking rigs by the burnt-out Sheraton Hotel in Kuwait City after the war has ended. (NH)

Journalist Lawrence Kiely undertook one of the first trips back into Kuwait after the war and reported his findings in *Truck* magazine, July 1991. Although he did not work for Astran, it is still interesting to read his comments as he would have seen exactly the same evidence of war that the Astran subbies saw:

'The roads were littered with debris and the place looked like a metal graveyard, with partially dismembered vehicles lying everywhere, twisted and broken. The place looked like it had been visited by some huge army of football hooligans, but could be repaired. ...

'A few drops of rain fell from the blackness above, forming small black dots on my companion's white t-shirt. By the end of that day, the spots had turned to holes. We were experiencing the horrible truth that mostly when it rains in Kuwait these days, it rains sulphuric acid. We wandered, coughing, among the parked vehicles. Even the newest were scarred by the corrosive precipitation which had been falling ever since Saddam fired the oilfields'.

Nigel Harness, one of the first drivers to enter Kuwait after the war, recalls a trailer load he took there for the rebuilding project:

'As soon as the war had finished my boss was happy to send us down again. There were three of us who took carpet and curtains for the new Sheraton Hotel which hadn't long been open. When we arrived we saw that everything had been burnt out. They had barely put the fires out and there we were with all the new fabrics and soft furnishings.

'I remember it was an American who was in charge of that particular site and as he hadn't seen our trucks he just told us to leave the containers where we could find space. It was quite funny as he didn't realise we had come all the way overland with tilt trailers, so after some deliberation the Americans found us somewhere to unload the stock.'

During this period a great deal of extra responsibility fell on Bob Paul's shoulders:

'I had a hell of a lot of respect for all the drivers who were running through the war and afterwards, and especially for John Harper who was caught up at the very start. I was extremely worried as to his whereabouts as I had tried to contact him but to no avail and I didn't have a clue what had happened to him.

'I knew he was in Iraq when things kicked off and then the next thing I had a telephone call from him from Saudi. He seemed very calm about everything. I can tell you I had a good few sleepless nights worrying about my boys and just praying that they were all okay.

'I just kept taking on the work because I knew our experienced drivers would find a way through, even if it meant them travelling the long way round. They knew I would look after them financially, so they were happy to do it and of course it did Astran the power of good with our clients who in turn were churning out products for their customers.

'What the drivers had to go through during the war was incredible and that period will always be etched in my memory as one of the most worrying times in all the years I ran the company.'

Before the Gulf War Astran had been delivering to Iraq for many years and since the early 1970s had used the country as a transit route to the southern Gulf states. However, after the war Iraq was virtually shut down, and trailer loads became few and far between.

None of the drivers seem to have had any real problems in Iraq in all the years that the company was going there. As Graham Wainwright said:

'The people were always the same. They lived under the same conditions before Saddam Hussein came to power and during his rule. They were all really nice to us. It was just the officials and the regime that caused the problems.'

Wainwright recalled a time when he helped an Iraqi national escape for a better life elsewhere:

'It was in the late 1960s when I met Adman who was married to an English woman from Reading. They were living in Mosul and hadn't seen many English trucks. Adman was a really friendly guy and he offered me the use of his house to have a shower and a meal. In the end I stayed the night, they were such lovely people.

'When I asked his wife what she missed most about England she replid, "HP sauce and baked beans." I just happened to have both in my truck so I gave them some as a way of saying thank you for their hospitality.'

Wainwright got on so well with the couple that they gave him their phone number to pass onto his colleagues so that if they were passing through the area, they could go and stay at the house. The arrangement went on for a year or so. Then, on a later trip Wainwright stayed at the house but noticed that the woman had gone. She had gone back to England. Adman had arranged for a telegram to be sent from Reading saying that her father was very ill and she was needed at home. She was then granted an exit visa and flew back.

At that time, no one was allowed to leave Iraq without a valid reason and when they were given permission they were only allowed to leave in the clothes they stood up in. Everything else went to the state. Wainwright continued:

'Adman told me he wanted to leave the country as well to be with his wife but he couldn't get permission, so I told him I would take him out. By this time I had done a few trips and knew how the borders worked so I was confident. I met Adman at the agreed place and he got into my truck carrying a small suitcase. I felt really sorry that he was leaving all his worldly possessions behind.

'We left Mosul and headed towards Baghdad then onto Kuwait. It was plain sailing but because I had a "passenger", I drove continually through the night and we arrived at the border at about 2 am.

'I told Adman to lie on the bunk because I knew the officials were too idle to come outside during the dead of night. I went into their nice warm office and threw a few packets of Marlboro on the desk. They didn't even look up at me, just stamped my paperwork and let me go.

'I drove on and then announced to Adman that we had arrived in Kuwait. I will never forget the expression on his face as he jumped around the cab. He was delighted. We drove into Kuwait city and booked in the Sahara Hotel.

'In his room he opened his suitcase and I was shocked at what I saw. It was stuffed full with hundred-dollar bills. He wanted to give me two thousand dollars but I told him no. He had always looked after me and the others and had always made us feel very welcome at his house. It was a lovely gesture but I refused his offer of the money.'

Adman had accumulated all the money after selling his tyre business, house and car to his brother. He flew back to Reading to be with his wife and then went back to university to study for a better life away from the Iraqi regime that he hated so much.

Twenty years later, Gavin Smith used the same route as Wainwright as he crossed Iraq on his way to the Gulf:

'By the time I was doing the overland run, the aim was to get through Iraq as quickly as possible. It wasn't like the olden days when the men had no urgency to get anywhere and customs formalities were reasonably simple. In my time, it was best to crack on because things were much more delicate and the longer you stayed in the country, the more chance there was of problems occurring.

'Rick Ellis, Mike Walker and Andrew Wilson Young, among others, would always do Iraq in one hit, driving straight to the Saudi Arabian border of Ar'ar without stopping. I have to confess, I generally chose a more sedate speed.

'My journey from Turkey to Baghdad was generally uneventful apart from the occasional roadside control. Normally they were customs checks, but almost everyone wore military uniforms so it was hard to tell who they were which made things a bit scary at times. They would all wear black leather bomber jackets and carry automatic weapons slung across their backs or sometimes clearly evident in their hands. Those guys somehow looked to be the genuine ones.

'There was one particular checkpoint that really did make me nervous. It was at Tikrit between Mosul and Baghdad, the birthplace of Saddam Hussein. At Tikrit control, the soldiers, customs, police, or whoever they were, were quite aggressive and intimidating and were always after some kind of souvenir.

'They had a knack of getting the driver round to the back of the trailer, supposedly to check that the customs seal was in place. While he was there, one of the guard's colleagues would be riffling through the cab – and there was nothing that could be done to stop them.

'An early departure from Mosul would ensure that I would reach somewhere like Samawah or Nasiriyah in the South of Iraq by nightfall. I had been warned that stopping at that end of the country for the night was a bit dodgy, because there were still some Iraqi soldiers who had deserted the army during the 1980s Iraq-Iran war. They were sometimes known to highjack drivers, although I hadn't heard of any Brits being done. As far as I was concerned, the quicker I was out of Iraq, the better.

'I never used to dwell too much over Iraq, but I do have fond memories of Safwan on the border with Kuwait. It is what everyone would imagine an Iraqi border to be like: dusty, ramshackle buildings built directly on the desert sand with nothing more than holes for windows and toilets and corrugated tin roofs.

'I always remember a particular woman official who stamped my documents, because she would always ask me for Western catalogues to see what the latest fashions were like.'

John Bruce remembers an incident in the 1980s when he was stopped by the military at an Iraqi checkpoint as he drove through the night from Basra into Kuwait. He had to do some very quick thinking to find out which soldiers had stopped him so he could be on their side when they asked the questions.

After the Gulf War, Astran was inundated with work for the construction industry, particularly for the rebuilding of Kuwait. The overland trailer services were running flat out and it meant that a lot of the subbies were turning round in Europe to take on another urgent load to the Gulf. Smith explained the system:

'Astran had so much work that it usually meant a quick turn-round once I got back to the UK, or more often than not, an empty 'flyer' from Doha to Rotterdam to pick up another load and go straight back down.

'For me, the priority was always Astran so those trips were great because they would pay us to run back empty at the rate that I would normally get for the backload from Turkey or Eastern Europe. It was worth doing and of course it kept me and the others very busy.'

To add their support to the British armed forces commitment in the Gulf, some of the subbies even had special t-shirts made to commemorate the ending of the war and the fact that Astran had kept the flag flying throughout the conflict.

Bob Paul was used to running his trucks through political conflicts and it wasn't just the Gulf War which caused him problems. As the political tensions escalated in the Balkans, the subbies found themselves regularly held up in massive queues of trucks all trying to enter and leave each country. Delays were very severe at times and added many days to the journey. Gavin Smith explained:

'It was considered normal to sit in queues all night long, not daring to sleep because you would be passed by several trucks whose drivers were itching for a chance to sneak past you and then hold back and let all their mates in. This would put the sleeper even further behind and was really frustrating, often resulting in arguments, fights and broken mirrors.'

The Astran Gulf War commemorative t-shirt modelled by Dave Buttons who is enjoying a celebratory meal of 'camion stew'. (BPO)

In the early 1990s a vicious war in the Balkans broke out, involving different groups in the former state of Yugoslavia.

Queues of traffic began forming not just at the borders but especially at filling stations. Because of fuel shortages some people were even pushing their cars to the pumps to save what little fuel they had.

Not wanting to waste an opportunity to make some extra cash, Gavin Smith and some of the other subbies found themselves able to help the desperate workers:

'We were in a position to assist the taxi drivers and farmers who were unable to work because they didn't have any diesel fuel. With the large-capacity fuel tanks on our trucks and trailers, we would cross through Hungary and fill everything up there at black-market rates of 300 litres for 100 Deutschmarks (DM).

'The selling price in Yugoslavia was 1-2 DM per litre, so there was a pretty good profit margin. The fuel tanks on my truck could hold 1,100 litres and the tank on my trailer could hold 2,000. I used about 400-500 litres to cross Yugoslavia, which left at least 2,500 litres to sell to the farmers and taxi drivers. Even at 1 DM per litre I made quite a few bob.

'I used to park at the Hotel National in Belgrade and phone a taxi driver who I had become friendly with over the years and who was trustworthy. He lived close to the hotel and would come to the car park to meet me. Then I would follow him down narrow tracks to his backyard where he had a huge tank and pump. It didn't take long to fill him up and I soon had a fistful of Deutschmarks.'

After Yugoslavia, the drivers entered Bulgaria where they could replenish their diesel supplies extremely cheaply. After completing customs formalities at the entry border, they had a relatively short drive to a dark parking area in the hills out of Sofia for their rendezvous:

'I would be met by two men in a very old, beaten-up Lada car with no seats in the back. Where the seats should have been there were drums full of Russian diesel. The rate for something between 400 and 600 litres was just 100 DM. It was a bargain.

'A few years later when they had made lots of Western money, the same men used to turn up in a huge tanker truck and blatantly pump the fuel straight into our tanks until it literally overflowed and poured onto the ground. I know it was naughty, but it was dog eat dog and we all had to survive.'

All the transit routes through Eastern Europe were now peppered with problems, so the drivers had to keep their wits about them to avoid the troubles as best they could. The more experienced men like Smith knew all the roads and were adept at changing their routes at a moment's notice. Smith described one such journey:

'I was running with another subbie, Ronny McNulty. After loading in Italy we drove into Slovenia without any problem which was amazing because by now it was the height of the war. We had to head towards Hungary via Ljubljana and Maribor because we were told that the motorway to Belgrade was blockaded at the Croatian/Serbian border and some Turkish trucks had been set on fire there. To make matters worse, one of the drivers had been killed.

'When we got to the Hungarian border we came across hundreds of buses full of evacuees, all trying to get out of the war zone. It was mayhem. There were vehicles and people everywhere.

'After jumping the queues – as was normal in those days if you could get away with it – and nearly having a fight with a tanker driver because of it, McNulty and I entered Hungary and headed across country towards Tompa. We parked for the night in the truckstop at Kiskunhalas where the superb liver soup and schnitzel washed down with some local beer were just what we needed after the long, hard slog to get there.

'Next morning we set off early to avoid traffic. Because we couldn't get out of Hungary into Serbia at Tompa due to fighting in the area, we decided to re-route and go eastward into Romania via the border at Nadlac and then head south, along the river Danube, then double-back into Serbia after paying the 'war tax'. It was then a short journey into Bulgaria through a tiny border at Bregovo.

'I will always remember one of the border guards there asking me, "Do you have cassette Beatles?" I thought it was a strange question, so I told him, "No, but my colleague has," and I sent him over to see McNulty who hadn't got a clue what was going on. I don't think he had any cassette Beatles either.

'Once we got into Bulgaria, it was a short drive to a place no bigger than a small village square where we parked up. The only problems occurred at night when the locals used to try and raid our trailer boxes and steal all the food, so we had to sleep with one eye open but at least we were clear of the war zone.

'After another early start, it was a long, slow punch right through Bulgaria to the border with Turkey at Kapikule to try and get through as quickly as possible. Then it was one hit to Istanbul where we arrived at the Oktay TIR park in the late evening, just in time for "Chok Efes Abi" ("Lots of Efes, mate"), Efes being the Turkish national beer. After all the hassles we endured through the troubled countries, a bit of R&R was definitely called for. From then on, it was usually plain sailing to Doha.'

Business for Astran continued successfully right through the 1990s. With his expert knowledge, Bob Paul managed to keep the transport operations running without too much hindrance, proving once again to his customers that Astran were indeed the experts on the Middle East run.

Throughout all the troubled times since the start of the first Gulf War, Jan Payne had been working away at Astran's HQ, dealing with all the documentation for the subbies and their loads. She gave an insight into Astran at this time:

'Before the Gulf War, our trucks were routed through Europe into Turkey, then via Iraq down to the Gulf. When the first war broke out it was obvious that this routing would not be possible and that the more expensive route via Syria and Jordan would be our only alternative. The additional expenses were due to the high level of transit taxes imposed by the two countries.

'At first we thought that our trailer services to the Gulf would have to stop because of those taxes and the risk of other Arab nations deciding to take part in the conflict. However, the war actually worked in our favour as the air-freight operators hiked up their prices to a ridiculous level and the sea-freight operators suspended all services into the Gulf. Therefore, the only viable transport was the overland road service.

'Fortunately for Astran all our subbies, bar one, were willing to run the gauntlet and as a result Astran had a very lucrative period.

As far as the Balkan conflicts were concerned, we didn't have any business from the war in Bosnia, but we did, however, have some work to Pristina in Kosovo. These were loads of materials to rebuild the power plants and were all under the supervision of a United Nations agency and had to be routed through Macedonia and then up to Skopje which was just short of the border into Kosovo. At this stage our trucks were not allowed into Kosovo.'

Bob Paul had seen his beloved company come through some difficult times over the years, including the first signs of financial trouble in the 1970s, the takeover in the '80s and then the Gulf wars in the '90s. Through all of those problems he was determined to keep the business afloat. Paul had built up a huge reputation for Astran and, with his tremendous passion for the job, was not prepared to let anything tarnish it.

Such was the respect that others in the transport industry had for Paul that he was invited become the Road Haulage Association's Middle East expert. While he was there, he was involved in the serious case of a British truck driver who was accused of smuggling drugs.

Paul's evidence in the driver's favour helped to acquit him. Paul felt passionately that the driver couldn't have known anything about the drugs because they were buried so deeply inside his load that he wouldn't have been able to

Bob Paul (front row, centre) at the surprise party held for his retirement. Paul was totally unaware of the event, but thrilled with the turn-out. (MT)

put them there himself. 'I was extremely pleased when the verdict was given,' Paul said. 'I will always remember the driver waving at me and shouting "Thank you, Sir," as he walked free.'

All seemed well at Astran, but as the new millennium began, Paul made the difficult decision to step down from the helm and retire in 2001 at the age of 71. Although he did continue with some consultancy work, he found it strange having nothing to do all day:

'To be frank, I was lost without the transport to run and it took a very long time to get used to it. Doing the consultancy kept me in touch with everyone because I used go the depot every few days which was nice – but eventually I gave that up too.'

Astran's surprise party for Paul came totally out of blue to him:

'I've never really been one for a big fuss and just wanted to go quietly but they were having none of it.

'I was absolutely taken in by the party. I knew nothing. The phone at home rang one day and my son Dominic answered it. He said a mate was having a do at the golf club in West Malling and we were welcome to go along. "God, I know that place," I replied. "It's right by my old depot. Any excuse for a drink. Let's go."

'When we got to the club, I clearly remember saying, "They've got an Astran sign up," but dismissing it as nothing unusual. One of my office workers' husbands worked in the bar and I thought it must be something to do with him. Stupid as it sounds, it still didn't click.

'As soon as I walked in, I was absolutely flabbergasted. The whole place cheered and applauded. There were all these people there to surprise me. I was thrilled and quite taken aback. It goes without saying that I had a marvellous time. Even though I wasn't bothered about a party to begin with, I soon changed my mind. I couldn't believe that it had been done specially for me.'

Would Bob Paul like to do just one more Middle East run, perhaps to his old haunt, Tehran? 'Not nowadays,' he sighed:

'It would spoil everything I remember about the job. The roads have changed so much and so have the trucks.

'When we started running to Tehran in 1966, we had to fight for everything as we went along which was what made the job so bloody hard but immensely satisfying when we got the other end, especially when the locals realised where we had come from.

'Look at the troubles there and across the Middle East these days. Anyway, I'm far too old to start dodging bullets now.'

CHAPTER 9

The Road Ahead

In October 2004 Uwe Ploog sold Astran to Hugh Thompson and Peter Carroll who each took 50 per cent.

Thompson and Carroll were both directors of Seymour Transport but, contrary to popular belief, Astran itself was never part of that business. It is true that there are some Seymour Transport trucks running around towing Astran trailers with the logo and lettering emblazoned along each side and across the back doors. This is simply publicity for Astran who paid for the sign-writing on the Seymour trailers.

After the sale, operations continued at Platts Industrial Estate for another year due to conditions in the property lease. That gave Thompson and Carroll enough time to make space for Astran at Seymour Transport's premises at Larkfield, Maidstone, so they could have everything under one roof.

The new owners wanted a fresh company without any previous liabilities. They registered it as Astran Cargo Services Limited, the name it is trading under today.

After the takeover, Carroll ran the Astran business while Thompson concentrated on Seymour Transport which he started over thirty years ago. Then there was a change in roles when Carroll bought Thompson out of Seymour Transport and at the same time took a smaller shareholding in Astran. Thompson was then in overall control of Astran and was assisted by his general manager, Kevin Letham, who had been with Astran since 2005.

The semi-retired Thompson has now stepped down from his directorship to concentrate on other business interests. He has made Letham a director of Astran to continue where he finished. Nevertheless, Thompson and Carroll are still the owners.

When Letham started working for Astran, he was

The Seymour-Astran connection is purely for advertising. (SL)

Celebrating the takeover. Left to right: Karen Vernon, Hugh Thompson, Bob Haffenden, Jan Payne, Kevin Letham. (ASTC)

despatched to Doha where he lived and worked for two months to gain an understanding of the business in the Gulf.

Astran has always maintained an office in Qatar. It is now manned by NP Ismail, an Indian who had been employed by Qatar Navigation for twenty years, where his responsibilities were to deal with the Astran trailers entering and leaving Qatar. For the previous twenty years the office had been run by Valerie Williams who retired when Astran was bought by Carroll and Thompson.

Since the 1970s Astran's main sponsors in Qatar have been Qatar Navigation, and they remain so today. As Letham explained:

'It is impossible to operate in an Arabic state without someone there who understands the rules and regulations and can act on your behalf. They deal with everything from car licences and insurance to other formalities which are far too complicated for anyone to understand. We couldn't be without them.'

When Letham returned from his initial visit to Doha, he began to build up the sales side of the business. In his time with the company he has overseen some impressive jobs including a concrete batching plant which was loaded in Italy and delivered directly to Doha. Although the contract originally called for three trailer loads, it ended up as twenty-six. In his time with the company Letham has seen some peaks and troughs:

'The batching plant was in 2005 which was a good year, but then 2006 was quiet.

'However, 2008 was probably the best year that Astran has ever had. We took 160 trailer loads for just one customer and in all probably did two hundred trailers to the Gulf states alone. I really don't think the old Astran did anything like that.

Astran Cargo Services Limited has been busier than ever in recent years. Pictured in 2005 in Qatar, some of the subbies with fully liveried rigs. Left to right: Roger Rabbit and two rigs owned by Graham Ball. (KL)

'Again we loaded in Italy for the Qatar Chemical Company which had made a major expansion to their plant. The main contractor for the work was Italian project forwarder Pan Projects who are part of the Panalpina group. We got the job of transporting everything which was very good for us.

'Before Christmas 2008 we had two low loaders go to the Gulf with valves packed in oversize cases for another client. The weight was only around 16 tons but it was the sizes which made it all interesting.'

Nowadays the majority of Astran's work comes out of Europe and in particular Italy and Spain:

'There is hardly any manufacturing left in the UK and in 2008 we only did eight trailer loads from here. Five of these were for a West London exhibitions company supplying equipment for Qatar National Day. The trucks stayed down there for two weeks and then brought the cargo back. The Middle East overland drivers will do anything so long as they can earn a shilling!

'This year, we are doing another exhibition job for the same client, for a sales seminar in Antalya in southern Turkey. Last year we also did three trailer loads of construction insulation down to Kuwait and a machine from Scotland which we took to Jordan but I really can't think of any more jobs we did from the UK.

'Concerning current political and religious troubles in the Middle East, the only recent problems we've had were in January 2009 when four British drivers coming home via Haifa, Israel, were caught when the missiles started flying over. They were all okay. I honestly don't think anyone will be coming back that way any more because the ferries from Haifa to Greece have just become very expensive. It's now around 2,000 Euros each way.'

The cost of each trip depends on the route the driver chooses to take. He must pay for his own ferry crossings, road tolls and transit taxes out of the money paid to him by Astran for the load which he has been contracted to deliver.

Nowadays, although Astran is solely a freight forwarder, the company still deals with plenty of overland trailer services and also sea and air cargoes. Agents are still present in all the major cities across

*A line-up of some of today's Astran subbies.
Left to right: Jeff Huddart, Terry Foggin, Steve Pooley, Andy 'Carrot' Blunsden and Chris Hooper. (KL)*

Europe and the Middle East working on behalf of Astran and there is one particular agent in Iran who will only deal with Astran in the UK. It is the same arrangement with the Turkish agent.

Letham said that it is Jan Payne who really runs the Middle East trailers and the Turkish service, while Karen Vernon operates the Iranian service. Sam Price helps Jan with Turkey and arranges all the sea-freight enquiries, while Letham himself keeps an eye on things. 'They are great girls,' Letham said. 'It's like being at home all the time. I get nagged at work as well as at home.'

Payne inherited the work which Bob Paul used to do before he retired in 2001, including sorting out the passports, visas and other documents for the vehicles and drivers. Payne said that most of the owner drivers now know how the Astran system works and will do their bit to help, although there are still one or two who rely heavily on her services:

'It's taken me a while to get used to them and I've had plenty of banter but I've got them all very well trained now. Bob Paul used to call me Mistress Jan and a lot of the men believed I was just that. I can assure you it was only a nickname but it kept them guessing. Maybe that's why they respect me so much.'

Bob 'Haffers' Haffenden started with Astran in 1977 and stayed with the company until March 2008 when he retired. He had specialised in dealing with imports for most of his time but in recent years Haffers has ended up dealing with the forwarding division to Iran:

'I was mainly responsible for organising a weekly groupage trailer to Iran. This was driven by Iranians or Turks who would come to our Maidstone depot to load and I'd have to deal with the papers and the language difficulties.

'I also got involved with Qatar General Petroleum (QGC). I dealt with their purchasing office in central London who would ring up and demand a truck that same day for an urgent load of pipes to be picked up from wherever and got to Qatar as quickly as possible. I'd do my upmost for them and arrange everything for the driver to make it as painless as possible for him. QGC had storage bays in Doha and if they became empty, it was a frantic race to get them restocked. Even though the pipes were probably not even needed, we had to keep the customer happy.'

'It was great to work for Astran as it certainly wasn't a normal nine-to-five job. I would say it wasn't ordinary. I went through a lot of the changes with the company and it took up a huge chunk of my working life but I enjoyed the challenge.

'It was more interesting when the drivers returned and told their stories but it went off the boil when the own-account transport shut down in 1989 as I never really saw any of them after that. Then it gradually changed from transport to general forwarding especially after Uwe Ploog sold it to Carroll and Thompson. Working there in the early days was certainly more interesting than of late.

'I was known as "Mr Bob" to the Iranian drivers who came to our depot. One day, two Iranians got caught coming into the country with drugs. The police found a note on them with Mr Bob and a phone number.

'I was shocked when my phone rang and a man asked to speak to Mr Bob in a very serious manner. He was from HM Customs and wanted to ask me a few questions. I got quite worried because I genuinely didn't know what he meant.

'The police and customs interviewed me, thinking that I was the destination for the drugs, so I had to explain to them what Mr Bob was all about. The driver ended up getting twelve years in prison.'

Jan Payne, operations department, has worked for Astran for thirty-four years although she did take some time off to run a pub with her husband. She returned to Astran six months later and called it her time off for good behaviour.

Margaret Stone, administration, started in 1985 as Uwe Ploog's personal assistant; she has now clocked up twenty-four years service.

Both women said that their best times were in the earlier days when the pressures weren't so great and things were more laid back. As Stone said, 'One day we may have stayed late to get documents finished, perhaps until seven or eight o'clock, but the next day, we'd come in later or just finish early and go to the pub.'

"We also used to have the occasional day-trip to France,' Payne added. 'That's something we can't do any more. We used to hire a minibus and driver so a load of us would go over for the booze. We miss those good days.'

When the drivers were employed directly by Astran, they kept their own trucks and it was easy for the girls to keep tabs on things. However, nowadays with many of the drivers dealing in sellers, it is slightly more difficult, as Payne said:

'They have a different truck every trip, so I have to keep my wits about me to remember which one they are out in. Sometimes one of the men will give me a wrong digit in the registration which throws a spanner in the works. I also have the pleasure of arranging all their documents including carnets, visas and passports. Although it's sometimes a lot of hassle, it actually is so much easier for us to do that. Everything is arranged for the drivers and then they pay us back afterwards.'

A driver who is particularly experienced in dealing with Middle East sellers is Barrie Barnes. Working directly for Astran during the 1970s and '80s, he then parted company and set up as an owner driver sub-contracting for Astran. Nowadays Barnes sources good second-hand trucks for his exclusive clients in the Gulf states:

'After the Gulf War I was finding times very hard as an owner driver and had virtually no money. After a particularly bad trip to the Gulf I decided to sell my truck to an Arab there who offered me good money for it. At the same time he asked me to bring him another one as soon as I could, and so it went on. He wanted another, then another.

'This soon became a good way to make money. I would find a suitable truck and trailer in the UK, then take on a load for Astran to coincide with delivering the truck to the Arab. I would deliver the load on the way down, take the empty truck and trailer to my client, collect the money and fly home.

'I reckon I've sold about two hundred although I've not delivered them all myself, because I've taken on drivers when I've been busy. I always make sure the trucks I sell are in tip-top condition because then I always get more work. A good trip will take sixteen days from when I walk out of my house until the day I walk back in.

'The routes and roads are very different from what they were in the 1970s. With modern building methods and technology, the road networks are as good as any in Europe. Nowadays we normally load in Europe, mainly Italy or Spain and then catch a ferry to Greece.

'From Italy, we use ones from Ancona, Bari or Brindisi, the latter two being the cheapest. The newest ferries are amazing and very quick. Supafast or Minoan can get to

Steve Pooley, one of the owner drivers now hauling for Astran, is parked by the Trans Arabian Pipeline in Saudi Arabia. (SP)

Patras in Greece in about twenty-two hours. The price for a one-way ticket is a little expensive at 1,000 Euros but for that I get a lovely air-conditioned cabin with TV, the use of an à la carte restaurant, swimming pool, disco, bar and a casino.

'Once I land in Patras I drive the 900 kilometres across Greece which takes about a day. Then it's into Turkey where the roads are now very good and it's more or less motorway and dual carriageway all the way through. It's around 1,400 kilometres to cross the country, taking about a day and a half.

'Although there is no tax to pay in Turkey, diesel is more expensive than in the UK. There is also a charge of around £25 to cross the Bosphorus Bridge. Once I get to Ankara, I always remember the age-old saying: the boys turn right for Saudi and the men go straight on for Iran and Afghanistan.

'When we arrive at Bab al-Hawa on the Turkish/Syrian border things become a little harder. Everyone gets charged to drive through Syria, calculated on the weight and value of the load and the number of kilometres. It costs around £400 to drive across at the moment. It's lucky that this is only 550 kilometres - otherwise it would be an incredibly expensive drive.

'For safety, trucks are convoyed from Bab al-Hawa overnight starting at 9 pm. They are taken to a holding compound close to Homs, and then once I get into Damascus it's a free-for-all. I've been going through the city for thirty-odd years – and there are still no signposts. I park up for a few hours sleep and then leave again in convoy at around 1 am and continue to the Jordanian border. Here there is mayhem!

'There are fourteen lanes of highway which filter into just one. Before I actually get into Jordan there are procedures to carry out and inspections which must be done at the border. It normally takes up a whole morning from six until noon. After that every vehicle has to drive through an x-ray machine which detects illegal substances.

'Finally, if all has gone well, I will get parked up around 3-4 pm.'

The convoy system was instituted in Syria after a driver took his truck into the centre of Damascus and blew up his vehicle and himself. Now all drivers are kept to the official route.

Entering Jordan also costs money and it is the most expensive country to drive through on the whole journey to the Gulf. There is no convoy system there, but instead the authorities sometimes fit tracking devices to the trucks so they can be monitored at any time. It costs around £400 to drive 175 kilometres across Jordan. 'I think it's outrageous considering it only takes two hours,' Barnes said.

'Saudi Arabia is relatively easy to negotiate compared with Syria and Jordan. First I must buy a set of transit plates at the border. These cost around £100 and include documents which the customs officials produce, showing details of the vehicle, the load, my name, destination, etc. There are police checks all over the place so all my documentation has to be spot on before I can leave the border.

'Saudi Arabia is around 2,000 kilometres long. The main highway follows the Trans Arabian Pipeline and at the southern end of it, everyone turns right for the border at Dhahran into Bahrain. If I'm lucky enough to get a load for Doha, it is then just half an hour from there to Qatar, my final destination.

'I will clear customs, unload the trailer, take the vehicle to my client and have a few days relaxing in a hotel before flying home. Then I'll start searching around again for another truck and trailer – and the whole routine starts all over again.'

A driver who has only recently started with Astran is Andy 'Carrot' Blunsden. In 2008 he did three round trips loaded to the Middle East and reloaded back to the UK. Although he'd had plenty of European experience, Carrot hadn't previously driven in the Middle East. He thoroughly enjoyed every one of his journeys:

'Jeff Huddart and veteran Chris Hooper took me on my first trip and on the second one it was me who took another new boy, Garry Cooper. and then the last trip I did I was with Garry Lyons and 'Brummie' Dave.

'The roads are almost perfect now and nothing like what I expected. There are plenty of proper truckstops and restaurants along the route although Damascus hasn't changed since I saw it on the film Destination Doha. I got lost there and I can tell you it was an utter nightmare trying to get back on route.

'Down in the Gulf, the people are in a different world. The rich are rich and it shows. I was amazed. I never came across any trouble on any of the trips. In fact it was the Saudi people who were the friendliest. They just wanted to shake our hands all the time and know which football teams we supported.

'There's no profit for an owner driver in these trips unless you can get a decent return load. And the quicker the trip is accomplished the better. I was lucky on my second trip as I got a load to Italy from the UK, then I reloaded in Italy for the Gulf and then ran empty back to Italy again for another load to the UK so that trip paid quite well. If I were to leave from the UK and go straight to the Gulf to unload, then run back to Italy for a reload I'd expect that to take around six weeks or so. It would be almost impossible to make any money on that.

'I can't fault Astran. I knew they were a reputable company back in the 1970s and it looks as though nothing has changed. All my documentation was okay, they paid me on time and I've got no complaints whatsoever about their helpfulness.'

So what does the road ahead hold for Astran?

Despite the economic crisis continuing in 2009, Kevin Letham seemed upbeat. He was confident that the amount of work coming would ensure a buoyant future:

'In March 2009 we were very busy with a new client working on a job from Spain to Qatar. We also had two English drivers, both of whom went to Mesaieed, Umm Sa'id, near Doha with the start of an order for forty trailer loads. We took on some loads for the Russian states, although most of that was sub-contracted to Eastern European hauliers.

'We are now one of the biggest trailer exporters to Iran and we are still sending at least one trailer load per week there and also to Istanbul. I think we did three trailers to Tehran and back in January 2009 the whole month was taken up with Iranian groupage.

'Our Russian work is also important. We became involved with a huge machinery job, our part of which was to deliver four trailer loads from the UK to Stupino near Moscow, via St Petersburg. There were also wide loads which formed part of the job too, but unfortunately they were too big for us to handle directly so were done by the Dutch heavy haulage specialists Van der Vlist.'

'Roger Rabbit' Gool was one of the four drivers:

'We took our rigs to Tilbury and loaded them onto the ferry for St Petersburg, then we had a few days off and caught a plane to meet the trucks at the other end. Then we had another week off stuck in the docks because there were some problems with customs formalities. We all stayed in a nice hotel so it wasn't so bad and once everything was sorted, we drove down to Moscow which took a couple of days.

Astran rigs in St Petersburg docks waiting for customs clearance. (RR)

Roger Rabbit poses by a Russian customs signpost, en route to Moscow. (RR)

No one knows what the road ahead has in store for Astran. Perhaps it will be the first overland journey to China.... (KL)

'After we had unloaded we tried to find return loads but Russia is such a bleak place none of us could get anything. So we had to drive empty back to Germany where we reloaded for England.'

Kevin Letham was keen to speculate where the next boom may be:

'Turkey seems to be the in place at the moment as Europe is almost finished and everyone is struggling to make any money there. Turkey it is the next stage on from Europe and business seems to be good.

'Our work continues to increase and we are still very busy with overland trailers. We had another cracking month in August 2009 with eighteen trailer loads for Doha from Corus Steel in South Wales. We also had ten loads from Italy to Saudi Arabia, four loads from Holland to Qatar and two from Italy to Doha.

'We have recently bought Lema International Shipping Ltd to add to our range of services. Lema specialise in Iceland, Malta and Cyprus, but it is all sea freight.

'Although we are never offered loads to India and China, I've heard that China is talking about joining the TIR convention. So you never know, the old silk roads that Michael Woodman and Bob Paul used all those years ago may open up again one day – which would be interesting indeed.'

How fitting it would be for Astran to secure the first load overland into China. Almost fifty years after Woodman and Paul set out on their epic journey to Afghanistan, another truck from the same company might be able to undertake another pioneering journey to a faraway place.

China would certainly not be beyond the ambition of current owner of Astran, Hugh Thompson:

'When Peter Carroll and I took control of Astran back in 2004 it would have been fair to say we had no idea of the depth of sentiment that goes with Astran.

'Certainly we knew of, and respected, the name. We knew the company had been a trail-blazer in the heady days of the '70s, going to far-off places that are normally only dreamed about. We also knew that times change ever more quickly, and the fickle world of road transport was not exempt from these influences.

'So at a time when analytical minds would have said otherwise, we ploughed ahead to raise the profile of this proud company. We concentrated on the core markets where Astran was so well known, and today are happy to report that Astran is in fine shape and able to secure business in the challenging times ahead.

'We will continue to drive Astran forward, but will never forget the efforts of all those before us who assisted in making the reputation of this outstanding company.'

CHAPTER 10

Astran Doha

In the 1970s when the oil and construction boom took off in the southern Gulf states, cargoes had to be delivered at a colossal rate to feed them. The cheapest way of getting freight to site was via sea containers and then local transport, but before long the Arabian ports became clogged up with ships waiting to unload.

To add to the mayhem, once the cargoes were finally unloaded, they were stock-piled on docksides for weeks or months waiting for import taxes to be paid. There was then the problem of arranging for the consignee to collect the cargo, if he ever did.

The more cargo there was on the dockside, the less chance there was of any room for the next ship to unload. So, many of the huge container vessels had no choice but to wait out at sea for weeks at a time before space became available in the harbours.

Saudi Arabia, Qatar and the United Arab Emirates (UAE) were all involved in a mammoth construction and redevelopment of their port areas. As money was no object, they began buying up everything they thought they would need and then they would buy everything that they didn't need. It was chaos!

Astran already had experience of providing a regular overland, freight forwarding and sea-container service to the Gulf states. Michael Woodman now spent time investigating the severe port delays. He concluded that he could distribute cargoes more effectively with his own trucks which would be more reliable than the local labour force and transport.

When Woodman sought a base for his trucks at the heart of the developing area, he chose Doha. At that time a small town, Doha had a naturally deep harbour which could easily be expanded and it was one of the places chosen for rapid port redevelopment.

The town was strategically placed to service routes to the north and south. Woodman thought that a base there would enable him to gain access to some of the more remote areas of the Gulf such as Oman.

With help from contacts made on previous trips, Woodman arranged for work to be available at the severely clogged-up Saudi Arabian west coast port at Jeddah and at Abu Dhabi in the UAE. To keep general costs down and use the Doha facility more effectively, Woodman thought that the overland trucks could be repaired and serviced there before they returned to the UK.

He set up a partnership with local customs agent, Intramas, who were based in Doha and whom Astran were already using for the clearance of overland services from the UK. Intramas would provide the work and an agreement was made for them to hold a 50 per cent interest in the enterprise. Woodman then found secure facilities for the men who would be based there and made an agreement with Gray Mackenzie, the Australian port management company, to share their depot near the port.

As stated in the previous chapter, it was vitally important to have a reputable local agent working on Astran's behalf. Before Astran could operate internally in Qatar, the company had to obtain the necessary licences and insurances. Woodman arranged for Qatar Navigation to supply these, which in effect made them the official sponsor for his company.

The depot was set up quickly. To begin with, one of the Astran overland trailers was delivered to the site and temporarily fitted out as 'luxury' accommodation and workshop. It was very crude but served a purpose. The front section was converted to a sleeping area and behind was a kitchen/diner/lounge area. A small air-conditioning unit was fixed up and a generator supplied all the power.

The first men to move in were John Frost, Geoff Frost (no relation), Kenny Searle and John Williams who was to be foreman and chief mechanic. They would spend three months working there, fly home for three weeks and then return for another three-month stint.

Geoff Frost has a distinctive memory of setting up the depot. 'We even had a brand-new Range Rover to get us about,' he said. 'And it came complete with huge shovel.'

Why a shovel? 'We had to nip out to the desert if we needed a big toilet job as there weren't proper facilities on site. The shovel was very handy. It was all mod-cons, you know.'

Even though they weren't strictly required to hold them, the men kept on the right side of the law by obtaining Qatar State driving licences. Geoff Frost thought it would be quite difficult and would require a test. In fact, all he had to do was take his passport and UK driving licence to

Working with Intramas, Astran also shared the depot with Kühne & Nagel. NVW 488P, driven by Roy Day, is loaded for the UK. (KB)

the police station, fill in a few forms – and within a couple of hours he had his licence.

The trucks based in Doha were taken from the overland fleet and were primarily 6 x 4 double-drive tractor units. These rugged trucks were much more suitable for the harsh desert conditions than the single-drive 4 x 2s, although there were also one or two of those. Two or three of the old drawbar outfits were based there as well. The Astran Doha trucks were kept legal by being re-registered with Qatar plates.

Kim Belcher, who spent six months at the Doha depot, drove one of the drawbar outfits which had come overland and was then left for the Doha boys to use:

'It was an old Scania V8, JAN 774K, which had had the bodywork taken off. It was useful for oversized or awkward loads that could only be craned off. Eventually a fifth wheel coupling, used to connect to standard 40-foot trailers, was bolted on over the axles.

'The truck was incredibly long and looked very strange. There was a huge gap between the back of the cab and the front of the trailer, but surprisingly it was quite comfortable and drove well.'

John Frost was one of the first men to move in at the Astran Doha depot. Note the generator and air-con unit fixed into the rear door of the 'luxury accommodation'. (KB)

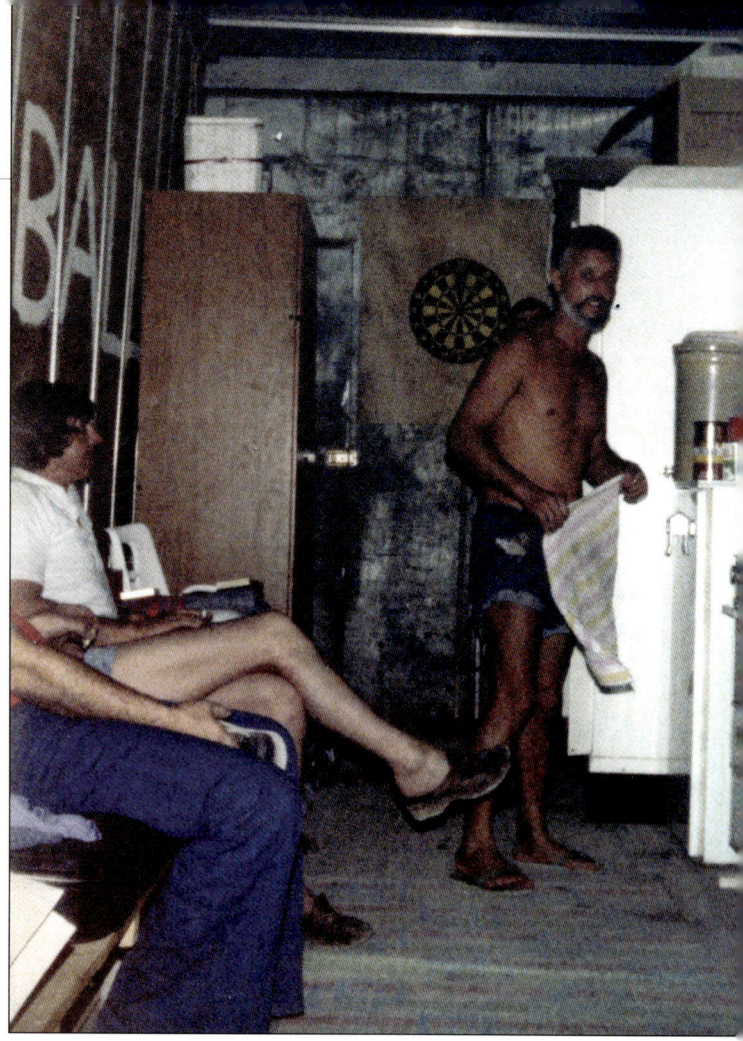

Living quarters inside the trailer were basic but bearable. Barrie Barnes (seated) and Vic Faulkner (non-Astran) enjoyed their time in Doha. (GF)

All the cabs were stripped out to a bare minimum to try and keep them as cool as possible although one or two did have air-conditioning units. PHK 172M, a Scania V8, was particularly favoured by those who drove it as it had heavy-duty axles rated for 100 tons and solid springs. It would go anywhere and was capable of pulling a proverbial house down.

Geoff Frost was its regular driver at Doha and remembers exchanging the normal road-going tyres for some gigantic sand tyres:

'The old girl would go across the desert without as much as a glitch on those tyres, but as soon as I took her onto the tarmac, they would instantly blow. You didn't want to be changing too many of those on a boiling hot day.'

JAN 774K was used at the Doha depot for internal work. It looks quite unusual without the normal overland container bodies. (TS)

NTW 562M, re-registered with Qatar plates, working internally in Doha. Just visible behind are PHK 172M and JAN 774K. (TS)

At the start of the Astran Doha operation, the main work was delivering oil pipes from Doha to another new dock complex being built further round the coast at Dubai. To speed up the process, the Astran men used to spend time at the docks helping to unload the ships. John Frost explains:

'Sometimes it took three or four days to get to our stuff, so one time while we were waiting, we helped some Danes out with a water treatment plant. They were really grateful and afterwards took us into Doha for a good night out. It was a great city and everyone got on so well. It was a really friendly place.'

The men got on well with all the expats living and working in Doha. Frost continued:

'They used to invite us to their clubs. A favourite place was at the Decca Navigation station from where all the ships moving around the Gulf were tracked. It was about fifteen miles out of Doha and many of the locals also used to frequent the place.

'We were always made to feel very welcome with plenty of food and drink. We had many a happy night there.'

The station was supposedly a top-security working environment but the employees were renowned for holding social nights. There was even a swimming pool.

It was not illegal to drink in Qatar so long as a liquor licence was obtained and, as Kim Belcher said, 'We had some really great parties in Doha. It was the best place to be at that time.'

Once the Astran Doha operation had gained momentum, Woodman flew John Griffiths, formerly Astran's UK warehouse manager, out to take overall control. Luckily for Griffiths, he stayed in a rented villa in Doha and worked from a nearby office. When John had done his share, he was replaced by Edward Dougherty who had previously driven trucks on the overland run for Astran.

As more work came in, Woodman replaced the accommodation and workshop trailer with Portakabins, much to everyone's delight. They were fully fitted out and included a decent shower and toilet. Each of the drivers had a cabin to himself which made life in the hot, dusty city much more comfortable.

Williams was pleased with the revamp to the workshop area. Originally there was nothing more than a tree and a container to which a block and tackle was

The Long Haul Pioneers

PHK 172M re-registered with Qatar plates and fitted with huge sand tyres to help with desert journeys. TDX 400K, alongside, still wears normal road-going tyres. (GF)

Kenny Searle poses with his 'beloved' Mercedes. Not the best truck for a tall guy to drive across the desert! (KS)

JAN 774K wearing Qatar plates. More importantly, note the Atkinson 'Silver Knight', just visible behind, which Peter Fanning drove to Dukhan. (KB)

attached. Now there was a proper overhead pulley installed to assist with major engine repairs.

As the port areas continued to expand, Astran Doha followed suit adding more trucks including a solitary low-powered Mercedes Benz tractor unit. However, the men thought this wasn't a patch on the bigger-engined Scanias and no one really understood why it was bought. Kenny Searle, an ex-army man standing over six feet tall, said that he struggled to sit in the Mercedes and was always banging his head on the low roof as he bounced across the desert!

Strange as it may seem, the Doha fleet also included an old British-built Atkinson 'Silver Knight' which had definitely seen out its working life in the UK. Quite why it was sent to Doha is another mystery, but according to Kim Belcher, it was Edward Dougherty who brought it in to be used as a depot shunter.

Nonetheless, Belcher said that it was also given more strenuous duties. More than once Peter Fanning drove it to Dukhan, a town about fifty miles across the desert to the west. 'Poor old Peter. No one else would drive the bloody thing. He certainly drew the short straw. It should never have left England.'

The bigger and more powerful Scania trucks more often than not pulled two trailers at a time to make the long desert journeys more profitable. Looking like Australian road trains, they were often loaded with two 40 ft containers which sometimes weighed around 50 tons each, making a total gross weight (truck, trailer, containers and loads) in the region of 120 tons … or more!

Other less weighty loads consisted of prefabricated buildings which were bulky rather than heavy, but just like the containers they were also trucked across the desert road-train style to maximise the truck's carrying potential. As Belcher said:

'No one could reverse the road trains. They just used to do their own thing. As the trailer wheels sank in the soft sand, the whole rig would end up in a zig-zag shape. It was quite funny, really. Mind you, we had a thousand miles of desert to turn round in so it didn't really matter.'

Geoff Frost laughed at his own memories of desert loads:

'Dynamite was one of the best. We used have petrol, fuses, caps, detonators and the dynamite all in one trailer which we took across the Empty Quarter of the desert to the geologists who were doing experiments out there.

'It was a bit bizarre and worrying. All the bits were separated in the trailer but that didn't stop the old bum twitching.'

In the early days there were no roads between Qatar and Dubai and it was about a hundred miles of desert

Some of the road-train loads were bulky rather than heavy. Originally this was PHK 172M, but she's now looking rather tired. (UNK)

Rick Ellis drove UAR 743S, here hooked up road-train style and ready to roll. (RE)

to negotiate. In the many roadless areas drivers simply followed the tyre tracks of those who had gone before. Because of the harsh terrain, some of the round trips to and from customers took a week. John Frost had one major piece of advice:

If you couldn't follow a Jordanian or another local driver across the desert, there was no point in going. There were no oil barrels or marker posts to guide you, so you could very easily get lost and then you would've been stuck for days before anyone found you … if you were lucky.'

One Astran driver – who shall remain nameless – did get lost when he became disoriented in the desert. He spent some time wandering around before the Bedouin nomads found him and took him to safety. His abandoned truck became victim to ferocious sandstorms. A helicopter had to be hired, at great expense, to search for the truck. Eventually it was spotted and a bulldozer was despatched – yet more expense – to dig it out.

The men worked the Doha trucks extremely hard. Considering that they had already done thousands upon thousands of miles service between the UK and the Middle East, they coped very well. There would be no need for them to be returned to the UK when their working life had expired or the depot closed down, so they were simply run into the ground and left in the desert.

With new construction projects starting every day, overland loads to the southern Gulf states were increasing dramatically. By 1979 on average two Astran trucks per week were leaving the UK on an express run to Doha. They would normally take a maximum of 10-15 days to arrive – but on some rare occasions, trailers were known to have been delivered and cleared customs quicker than air freight!

Dick Snow, Charlie Norton and Dave Poulton were three men who did the overland run regularly … and fast. They did not want to spend a long time working at the Doha depot, but Snow and Norton were not averse to helping out if the need arose. Norton explained:

'We'd arrive at the customs in Doha with a loaded trailer from the UK and while it was being cleared and unloaded, we'd do some work to help the Doha boys out.

'I went across to Riyadh in Saudi Arabia and back which took about a week. When I got back, my original trailer had been unloaded and serviced for me so I'd be on my way

Barrie Barnes drove KJN 671P and would help the Doha boys when necessary. Coupled to two 40 ft heavy containers, the rig grossed over 100 tons. (BB)

again soon after. It was better than sitting around for a week in the baking desert heat. Payment for the extra work was good – and it was always cash in hand.'

John Bruce was another overland driver who would volunteer to help out:

'I would go round to Abu Dhabi for a load of fresh water. It would take about a day but it was a necessity. It was a very important routine that had to be done.'

Apart from delivering trailer loads, there were other jobs which Astran Doha had been contracted to do. Preparing imported trucks was one example. The Arabs were flush with money but had absolutely no idea what they needed or what they were buying. They were so desperate for trucks to move their cargoes that they would buy anything they could get their hands on, even vehicles which were not remotely connected with desert work. Norton remembers working on a couple of Canadian trucks:

'They had been imported direct from the oilfields in the very north of their country and were fully "padded up" with all the cold-weather gear that they needed in the extreme areas where they belonged. It was so funny to see these trucks

Perks of the job. Kim Belcher bought this custom-built 7.5 litre Corvette while he was working in Doha. (KB)

arrive off the ships with quilted radiator grille covers and electric blankets wrapped tightly around the fuel tanks. It was plus 40 degrees where we were!'

All jobs have perks and the Doha boys had their fair share of them. Kim Belcher recalls one instance which made him the envy of all the others:

'I'd done some regular deliveries to a wealthy Arab and I'd noticed he had a very impressive American Corvette car which just seemed to be collecting dust. I loved my cars at the time, so I asked him about it and was told the car wouldn't move as it had problems with the gears. It was a fantastic-looking motor and, being young, I really fancied it for myself, so I struck a deal with the guy and bought it off him for a song.

'I towed it back to the depot and got John Williams to have a look at it. He got underneath and fumbled around for a while then used a pair of pliers to simply un-kink a little mechanical device which ensured the car went into reverse gear whenever the ignition key was removed. The gear problem was resolved. It was as simple as that. I was over the moon.'

Belcher's new car was a top-of-the-range motor imported directly from America with a unique spec. It had a 470 cubic inch engine giving it something like 7.5 litres of power. The extras included a pair of huge stainless steel side-pipe exhausts! Belcher was elated when he managed to get the car going and spent all his spare time driving it around Doha.

However, when Belcher left Doha he was unable to obtain the permits to ship the Corvette back to the UK. Very grudgingly he had to leave it at the depot where it collected dust until he sold it onto another Arab. Belcher does admit to having made a tidy profit on the deal.

Considering the amount of construction work in progress in the southern Gulf states, it is surprising that the Astran operations at Doha were not as successful as they should have been. Michael Woodman had employed a couple of sales people to work in Doha, but unfortunately they didn't bring in much new business.

Tony Soameson was then asked to move there for six months to try and expand the business. He shared the villa with Edward Dougherty and recalled a favourite pastime: 'Apart from watching TV, we used to shoot cockroaches which were running along the walls. We had air rifles and it was great fun as some of them used to go with a right crack.'

The 'white elephant' barge service from Rotterdam to Saudi Arabia. Around two hundred trailers could be carried but the journey was extremely slow. (TS)

Soameson knew his sales job well and managed to get more work from local Arab companies, liaising with Intramas and the men at the depot. Although he didn't get involved with the actual running of the trucks, Soameson did play an important part in the Astran Doha operation:

'One of the things I tried out was a barge service that came into the port of Yanbu' al Bahr, north of Jeddah. Unaccompanied trailers were loaded onto it at Rotterdam and the whole thing was then towed by tugs all the way to the Gulf. It was the biggest white elephant I've ever known because when the seas were rough the flat-bottomed barge used to roll terribly and make the trailers very unstable. Much of the cargo was damaged. It was also very slow and just wasn't feasible for us.'

Although the barge service was nothing to do with Astran directly, Woodman did charter space on it to see if it was a cheaper and quicker way of getting the containers and unaccompanied freight to the Middle East. The trucks based at Doha travelled to Yanbu' al Bahr to collect the trailers but organisational problems at the port didn't help the operation. To make things worse, Soameson said, the local agent in Yanbu' al Bahr wasn't the best, either:

'He worked from his house and regularly had hundreds of documents spread all around him on the floor with various other people sitting cross-legged drinking tea and smoking their hubba bubba pipes. I remember walking into the room once and being shocked at what I saw. No wonder nothing ever got done.'

The agent also struggled to translate the details on the manifests from English into Arabic. The whole procedure of getting customs clearance for the freight took weeks at a time which was totally unacceptable to the business. Soameson told me that he oversaw two or three of the barge trips but had to give the idea up as a total waste of time and money.

Other financial problems finally contributed to the demise of Astran's operation in Doha, so after two or three years it was decided to close everything down and return the employees back to overland duties.

Intramas continued to act on behalf of Astran as it always had done, dealing with import and export freight of overland trucks and sea containers. The company is still to be found in Doha today continuing with the very healthy business connected to Astran Cargo Services Limited.

The Astran Doha depot. SOO 815R, driven by Terry Tott, and NVW 488P, driven by Roy Day, are loading for the UK. (KB)

Roy Day bids a fond farewell to his mates at the Doha depot as he steers NVW 488P back to the UK with a stricken Gulf Air helicopter. (KB)

CHAPTER 11

Turkish Delight

'Turkey was definitely the most challenging and most exciting country to get through,' Gordon Pearce said. The half-way point to the Middle East, it held many hazards for the men who pioneered the routes. 'The people themselves were very friendly, it was just the terrain which was formidable and unpredictable,' Pearce added.

Nowadays, the Turkish road network is excellent with the majority being either dual carriageway or motorway making for a reasonably quick journey through the country. With good communications and proper infrastructure in place, Turkey has now become as good as any other Western country.

However, in the late 1960s when the first Asian Transport trucks were heading east towards Iran, most of the roads were little more than dirt tracks. It would have taken the men a good four or five days – if they were lucky – to cross the country, all 1,500 miles of it, from Kapikule on the Bulgarian border to Bazargan on the border with Iran. Pearce continues:

'The adventure has definitely gone now because the roads are so good and trucks are so technically advanced and powerful, there is no fun any more. In the early days we used to pass through every single town and village on the way through Turkey and it was quite a highlight for the locals to see us in our brightly painted trucks complete with strange words such as "London" written on them.

'Even the so-called main roads were more often than not quite narrow with buildings tightly packed on either side, but with little traffic in those days, we could sometimes drive through villages at full speed, 40 mph.

'We used to compete to see how many caps we could knock off the men's heads as we whisked past. There were always people milling around and as they saw us approaching they would stand right on the edge of the paths to watch, so it was too good an opportunity to miss. We could see them in the mirrors shaking their fists at us as they bent down to pick up their caps.'

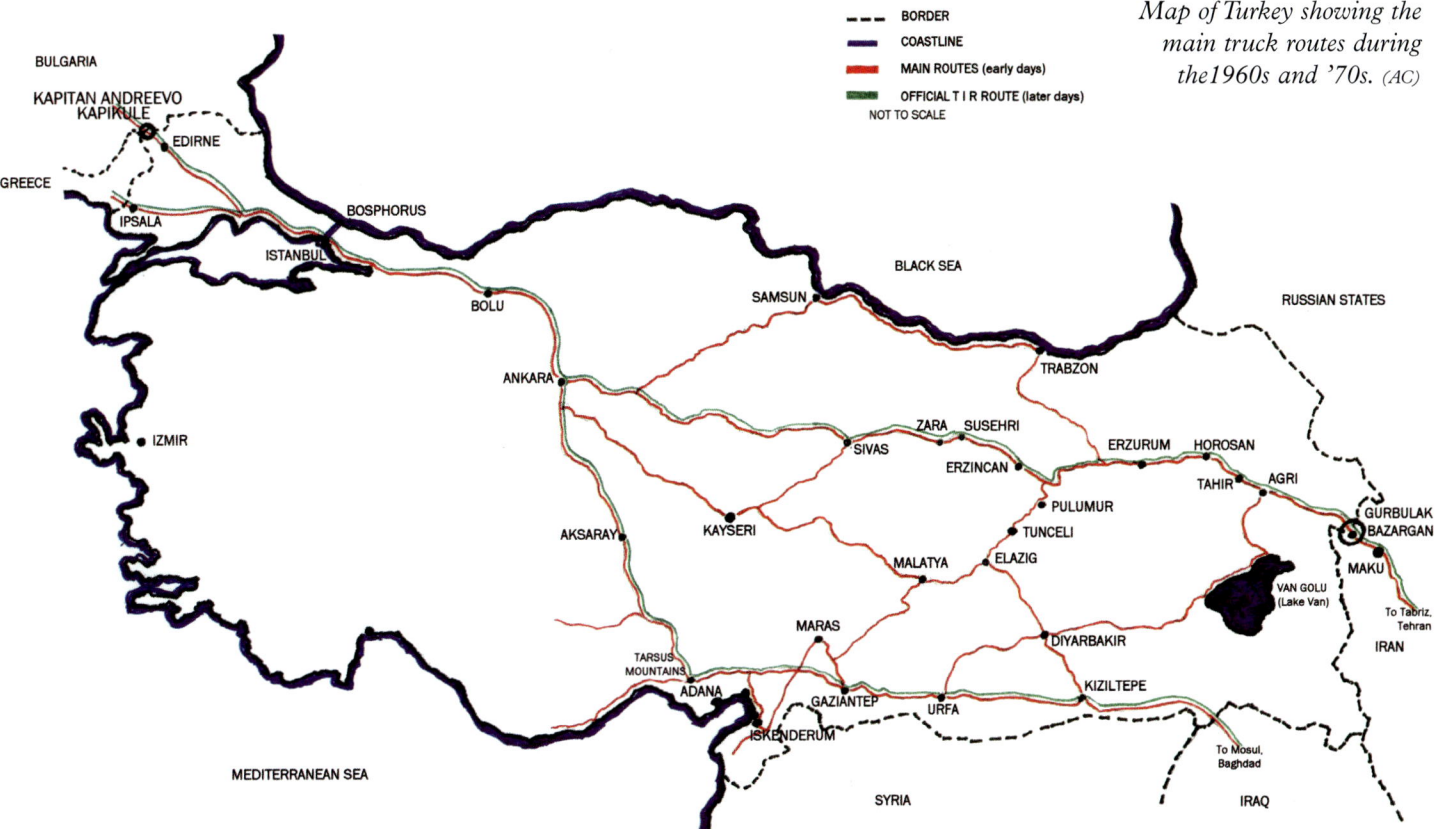

Map of Turkey showing the main truck routes during the 1960s and '70s. (AC)

213

Gordon Pearce steers UHM 25F through a Turkish village. Note the men with caps on standing to the right. (ASTC)

'We were proper adventurers and did the trips by the seats of our pants,' said Graham Wainwright. 'No two trips across Turkey were the same and there was always something different to cope with each time – and all for just twenty quid a week.'

Peter Cannon got to know the country very well during his time as a driver:

'If Turkey hadn't been there it would have been a boring trip. The country was interesting to say the least and full of historical sites. It was very picturesque, awe-inspiring and the people were always helpful. Wherever we stopped they were inquisitive, interested and pleasant. To drive across Turkey was very challenging which at times made it difficult and very dangerous, especially in the bitterly cold winter months.'

In this country of great extremes, Western Turkey has always been civilised with fairly flat, good roads and large towns and cities. It is a popular place to live, with generally pleasant weather.

On the other hand, Eastern Turkey, which constitutes the bulk of the country, has always been much wilder, with very high mountains and lots of snow and ice in the winter. There has been constant migration from this inhospitable land to the kinder area in the west. In the early years, the rough, unmade tracks in the east had to be graded regularly throughout the year with large bulldozers and motor scrapers to keep them in some sort of order and to make them passable.

In the mountains, mainly in the east, the roads were steep and twisting with bad cambers on the sharp hairpin bends. Some of them were so bad that if trucks were heavily loaded, they had to drive around the hairpins on the outside edge – the wrong side of the road – where it was not quite so steep. Any traffic coming from the opposite direction would have to pass on their wrong side which obviously took a bit of getting used to.

In the winter it was essential to fit snow chains on the drive axles of the truck and often another set of chains on the front steering axle if the weather was very bad. Cannon said that Turkey was very challenging:

'We would even put one or two sets on the rear trailer axles if the weather got exceptionally bad and I can admit to being quite scared on several occasions. Imagine a 40-ton truck and trailer coming down a very steep mountain road in deep snow. There is a blizzard at full rage and it is so dark you really can't see very far, if at all, and you are very aware that there is a steep drop directly at the side of the road with no safety barriers to protect you.

'Worst of all, the weight of the outfit is pushing itself down the mountain at a speed faster than you can control!'

But before tackling the delights of Turkey, the men came through Bulgaria. Arriving at the country's exit border at Kapitan Andreevo, they completed the TIR documents and drove a short distance through no-man's-land to the Turkish entry border point of Kapikule.

In later years, through the 1970s and '80s, Kapikule was to become one of the most dreaded and inhumane places that a Middle East truck driver would ever have to encounter. If there was rain, trucks would literally become bogged down where they stood. Due to the inefficient authorities and the extremely slow rate at which trucks were processed, there were times when queues would stretch fifteen or twenty kilometres back into Bulgaria. It would sometimes take four or five days to get into Turkey simply due to the huge volume of traffic trying to get through such a small border post.

Today, 2009, Kapikule has become the second busiest land crossing in the world. In 2008, 400,000 vehicles and four million people went through it.

However, in the mid to late 1960s Kapikule was little more than a very quiet, simple border control, as Peter Cannon recalls:

'When I went through for the first few times, there were never many trucks, just us and a few Bulgarians and Romanians. They had some old Leyland Ergomatic cabbed tractors which were all in very bad condition. I always remember the drivers had nothing and were especially short of food. They had no spares and definitely no money.

'The main traffic at the border was Turkish private cars full to the gunwales with gastarbeiter, Turkish guest workers, travelling to Germany to work in the car factories. Later during the summer, I'd see them all returning to Turkey driving second-hand German trucks loaded with goodies. There was usually a small queue of them at the border consisting of a big truck, carrying two or three smaller ones.

'It was quite comical to stand and watch the Turkish drivers trying to do a deal with the customs officials in order to import them into Turkey. There was always a very high rate of duty to pay on vehicles and I suspect quite a lot of money changed hands illegally, rather than via the official method of paying which would have cost an awful lot more.'

It was essential to fit snow chains to drive axles and front steering axles to get maximum possible traction. (JF)

Graham Wainwright also has memories of Kapikule:

'There was only room for about a dozen trucks. There were a bar and a small motel where we would usually stay for about twenty-four hours while the officials checked our documents. If we needed anything translated, or there was a language problem, there was a good interpreter who based himself in the hotel and we got to know him very well which helped a lot. It was all very civilised and friendly in those days, but then again, there was only a handful of us.'

Once the men were in Turkey, the next stop was Istanbul where they would head for the BP Mocamp truckstop located on the western outskirts of the city near the airport. Just about everyone stopped there for some well-deserved time off. Bob Paul remembered it well:

'Michael Woodman and I stopped there on our first trip to Afghanistan in 1964. It was just a very small place with a garage and a restaurant. While we were there, we met lots of Australians and English people in their jeeps and camper vans. It was a great meeting place for all the backpackers and travellers so we stayed there for a while and had a really good time with everyone. They were all amazed to see us in our little truck.

'In the early years of Asian Transport, my drivers and I always made a point of stopping there as it marked more or less the half-way point in the journey. There were only us to begin with.

'Then as the routes got busier with more and more trucks, the owners realised there was profit to be made from the passing trade so they began to build the place up. They bought a plot of waste ground to the side of the main property and turned it into a huge truck park. It quickly became a haven for drivers of all nationalities. When it got really busy in the '70s I had virtually to ban my drivers from stopping there because some of them were spending far too long socialising.'

Gordon Pearce recalled the Mocamp being spic and span:

'Bob Paul spoke highly of the place and when I arrived there on my first trip, I was quite surprised at how small it was. We had to park on the garage forecourt because there was no parking area at all. I remember the owner used to watch us parking because he had painted all the kerbstones white and he was very insistent that no one touched them with their tyres.

'At the fuel pumps we were even greeted and served by very smartly dressed pump attendants. Everyone kept the whole place immaculate.'

Normally a couple of days would be taken at the Mocamp. Drivers would don their overalls and spend a day servicing and greasing up their trucks and trailers,

Bobby Vallas outside the bar at Kapikule in about 1969. Behind are the motel and parking area. Note the Bulgarian Volvo F88s. (BV)

checking brakes and tyres and doing general maintenance. Once they had done their duties, had a good feed and been thoroughly watered, they would depart, leaving Europe behind to start the really hard drive into Asia.

Ray Scutts remembers the Mocamp for a very good reason:

'When Gordon Pearce and I arrived there on our way back to England, Gordon had a letter waiting for him from

Gordon Pearce, driving BGH 172H, arrives at the Mocamp Istanbul. (GP)

216

The Mocamp was popular with backpackers and truckers alike. Gordon Pearce (far left), Bobby Vallas (3rd from left) and Bun Parlane (far right) with some of the travelers in about 1970. (GP)

his wife. My wife and his were friends and the letter said I had become a father again and this time I had a beautiful daughter. I was so thrilled, but it was no good celebrating in Turkey with Coca Cola, so we drove home non-stop and arrived at the docks in Holland just five days after leaving Istanbul. That was some going in those days.'

Spending some rest time at the Mocamp was all part of the job and Graham Wainwright remembers passing two or three days there with Bob Paul before they departed for Asia:

'We wanted to have a proper look around Istanbul so we stayed in a nice hotel right in the central area. I loved the city and used to go to the markets to buy suede coats and leather trousers which I sold in Holland on the way home. I think I paid about £15 for a really nice top coat made of soft, luxurious suede. The Dutchmen loved those sorts of clothes which went nicely with their clogs and wacky baccy.

'I really enjoyed the Turkish food. The spicy smells and wonderful colours of the ingredients were so unusual and I will always remember the breakfasts we had at the hotel. The chef threw everything in a big pan: tomatoes, mushrooms and eggs, then cooked it for a short time in fragrant oil. The eggs were never cooked thoroughly, though. They were always snotty but we used to mix it all up with fresh bread and it tasted lovely.

'We used to go into some of the local restaurants and ask what they had to eat and could we see it before we ate it. The chef would lift the lids off the containers and then sweep the flies off. It was all very appealing. There was always a saucer on the table with herbal grass piled on it. It was there to eat with the food.

'I used to love the Turkish natural yoghurt which I enjoyed together with flat breads and we used to roll the grass into the bread which stopped us getting the trots. It seemed quite effective.'

Two drivers who were often paired up in the early days were Bobby Vallas and Dick Snow. They shared a truck for about two years and became very well known on the route to and from Iran but were especially famous throughout Turkey. One big man and one small man, they were quite an attraction wherever they stopped, as Vallas recalled:

'I remember a restaurant in Istanbul where we once met a famous local wrestler. We had a few drinks with him and then he challenged us to an arm-wrestling match. He saw that I was short and thought he could beat me. I'm afraid he didn't because I had very strong forearms.

'From then on, every time we used to get near the place, the locals knew we were on the way and it used to get packed. They would shout "Bob's coming!" and the wrestler would come and find us. I used to think, "Oh no, here we go again."

'Dick and I always used to win, but they never seemed to mind. Then one day I said to Dick, "We can't keep winning. It'll cause trouble, so why don't you let him win tonight." After a few struggles and with lots of grunting and moaning, Snowy finally gave up and conceded defeat. The man, delirious with his victory, ran out into the street shouting and waving to anyone who would listen. The celebrations went on long into the night – and Snowy and I just had to join in.'

The Long Haul Pioneers

Before the multi-lane Bosphorus Bridge was opened in November 1973, a ferry was the only way to cross from the European west side of Istanbul to the Asian east. As Gordon Pearce recalled:

'We used to leave the Mocamp at about 6 am to miss the queues at the ferry terminal. The quayside at Sirkeci was wide enough for a line of trucks to park while they waited for the ferries. Most of them were Tonkas which were always grossly overloaded and I was surprised that the boat never sank as it had so much weight on it.

'Even getting to the terminal so early, we still had to wait a couple of hours before we could board but once we'd done it, the crossing was only about twenty minutes. There was only one deck on the ferry and we could just about squeeze two of the drawbar outfits on, one behind the other. It was very tight to get vehicles on board and it nearly always ended up with one or two trucks touching each other.

'There was a small snack bar upstairs where we could get a quick cup of chai tea and a sandwich. We paid cash for each trip on the ferry. I think it cost about 200 Turkish lira each way which was very expensive in those days.'

Peter Cannon recalled the procedure for the ferry and carnets that were needed for the temporary importation of the truck and trailer (nothing to do with the cargo) into a country:

'To get there we kept on the main Istanbul road until we came to the old city walls which were very high. We used to turn right there and follow that road outside the walls until we came to the sea. Then we'd turn left and follow the coast road until we arrived at the back of the queue for the ferry which was actually in the old city, quite near the Galata Bridge.

'Every time I went on the ferries, there was a queue and the worst bit about it was

Gordon Pearce, driving UHM 25F, is greeted by a friendly fuel pump attendant at the Mocamp in 1968-9. (JD)

The winning team: Dick Snow and Bobby Vallas, holding beers, celebrate another arm-wrestling win with locals. (RB)

218

that we could never sleep while we waited simply because we'd get overtaken. We had to take our carnet de passage documents, one for the truck and one for the trailer, to the ticket office in order to buy a ticket. They took the unladen weight and the registration number of the vehicle and charged us accordingly.

'The Bosphorus was a very busy stretch of water at all times. As we sailed across and looked at the magnificent sights of Istanbul, I always thought to myself, "Now the trip really starts. I'm in Asia at last."'

The next major town on the route through Turkey was Ankara, a good day's drive which included stretches of road which needed extreme caution to negotiate. One such section was a long, slow climb up and over the 5,000 ft mountain pass at Bolu and another was a dangerous decent known as Death Valley further on between Bolu and Ankara. On this steep, twisting road, with a rock face on the left and a gorge on the right, some drivers lost control of their trucks and some even died – though not Astran men.

Trying to slow down forty or more tons of truck purely by braking heavily would very quickly cause 'brake fade'. The brakes would overheat and catch fire. The trucks would get faster and faster and there would be no way to stop them!

When the drivers finally made it to Ankara some headed straight for the British Embassy in the centre of the Turkish capital. Pearce recalls taking deliveries of desperately needed supplies for the Consul there: 'I regularly used to take him Heinz tomato sauce and Kellogg's cornflakes, his favourites but impossible to get hold of in Turkey.'

There was also a good stopping place for drivers on the outskirts of Ankara. A motel near to the Turkish military airbase provided the men with proper facilities such as hot showers and Westernised toilets, which everyone appreciated. 'It meant we didn't have to squat over a hole in the floor for a dump. It was the last luxury before the going got tough,' Wainwright said.

At Ankara the men had to decide which route they would take across Turkey. That depended on the time of year and more importantly, the weather conditions. There were four main routes available in the early days and the decision of which one to take was helped by the fact that there were very few restrictions for commercial vehicles.

John Williams, driving JLL 686K, boards the Bosphorus ferry on the west side of Istanbul. Subbie J&T follows closely behind. (ASTC)

Just enough room for everyone to squeeze onto the ferry. (ASTC)

Turkey later prescribed a route which all TIR vehicles had to follow. It was an attempt to prevent them going through all the towns and villages and was a way of keeping check on them across the country. 'By then it became quite boring following all the other hundreds of trucks,' Pearce said. 'I used to enjoy the driving much more when we could vary the routes.'

Tony Soameson, driving AMY 147H, negotiates the terrible section of road between Zara and Susehri near Sivas in April 1970. (PC)

The route which caused most problems in the early days was the one via Sivas and Erzurum, taking a more or less central route eastward across the country towards Iran. The main problems were caused by very steep roads between Zara and Susehri to the east of Sivas. Great care had to be taken negotiating the many bends and cambers because some of them were so bad they would force the front of the trailer to hit the back of the truck causing the roof to become damaged or even ripped open. As Cannon said:

'The reason that we had to look at the different routes was because of this notorious short section of road. It was quite high in altitude, too, and in the winter there were sometimes six feet of snow which made it totally impassable.'

Sivas itself was also a major trouble spot mainly due to the extremely steep hill leading down to the town. Its gradient was around one in five and it was cobbled which made it very dangerous when it was wet. If that wasn't bad enough, the approach road, if it could be called that, was also extremely steep. It was virtually solid clay which again made it incredibly slippery, especially when it was wet. During wintertime, the roads were totally impassable, as Pearce desribed:

'The Iranian and Turkish drivers had told us scary stories about Sivas during the winter, where trucks had run out of control and crashed through buildings so that was enough to put us off. If it was definitely dry weather we would go through Sivas, but only if we didn't have much weight on.

'In wintertime we'd divert south to the coast as the climate was warmer but it took a lot longer going that way. From Adana we went east to Urfa, Kiziltepe, Diyarbakir and then via Van Gölü (Lake Van). If it was warm enough, we'd stop there for a swim and a meal on the shores. It was so tranquil even though it was a military road which meant we sometimes had visits from soldiers who were patrolling the area.'

The approach to Sivas. Very twisting and made from clay, it was extremely slippery when wet. Note this is a two-way road. (GP)

Looking down the steep, cobbled hill into the town of Sivas. (GP)

Joe Bowser, driving GUL 588N, experienced a sudden snow storm in May 1977 which stopped him in his tracks on the approach to Sivas. (JBO)

Tony Soameson, sharing YYW 330G with Peter Cannon, takes a dip at Lake Van watched by a couple of Turkish soldiers who were patrolling the area. (PC)

The view towards Malatya had some breathtaking scenery. The Vabis, UHM 26F, in the far distance puts into perspective the vastness of the countryside. (GW)

In springtime the road resembled nothing better than a mud bath! Trucks often got bogged down and needed towing out. Bun Parlane surveys the situation. (BPA)

The trailer, now waiting to be pulled to safety, was trying to slide away in the mud. Even the grader had chains fitted. (BPA)

An alternative route from Ankara went north-eastward to Samsun on the Black Sea and then followed the beautiful coastline eastward to Trabzon, rejoining the main road near Erzurum. This, too, was a long route which included some nasty climbs and cambers on the last stage between Trabzon and Erzurum which would take a day's driving alone. Due to the twisting mountainous roads, it was virtually impossible to get any decent speed up which also hindered the men.

In summertime, the most favoured route from Ankara was to follow the road south-eastward to Kayseri and Malatya, then north-eastward through Elazig and Tunceli, joining the main road near Erzincan. The section near Malatya was spectacular with some breathtaking scenery.

It may have been an idyllic route in the summer but Graham Wainwright said that in springtime the road north of Elazig was always covered in thick, glue-like mud which was as bad as being stuck on ice and snow:

'I got stuck badly in the Scania Vabis demonstrator once. The complete rig got bogged down up to its axles and the trailer was trying its hardest to slide all over the place in the mud as the truck's wheels struggled to get any sort of traction. The whole rig was zigzagging all over the place.

'The cab on that truck was tiny and my mate Bun Parlane and I got absolutely plastered. We were up to our waists in mud and the cab was in a right mess too.

A typical gang of council workers, karayolari, whose job it was to keep roads clear and tow stricken rigs to safety on the mountain passes throughout Turkey. (GW)

'Another time, I was accompanied by Bun Parlane again and we were going outward to Iran in UHM 26F, a different Scania Vabis. We only just about managed to get to the top of the pass with the help of the local karayolari (highway workers) with their motor graders and bulldozers. I wondered if we should have gone that way in the first

Graham Wainwright, driving UHM 26F, gingerly pushes the trailer onto the soft mud as he tries to turn the rig around. Bun Parlane is helping. (GW)

place, but by the time we got there it was too late to change our minds.

'We were due to rejoin the main road to Erzincan on the other side of the pass, but before we even got over, one of the bridges had been washed away. There was nowhere to turn the 60-foot drawbar outfit around so I had no choice but to reverse the whole rig backwards which I knew would take hours. With Bun directing me from behind we made very slow progress.

'Then luckily I found a wider section of dirt on to which there was just enough room to push the trailer. It was very soft, but I decided to try and turn round there. Being heavier than the trailer, the truck would have certainly got bogged down, so I unhitched, left the trailer where it was with Bun guarding it and then drove forwards again until I found a suitable place to turn the truck round.

'Then I came back and hooked up to the trailer using the front bumper hitch on the truck. At least I was facing forwards now and, with Bun's guidance, I pushed the trailer like that for the rest of the way. It wasn't ideal and took an hour or so, but at least I could see where the edge of the road was which made it a lot easier than reversing.

'We persevered and eventually found enough room to turn the whole rig around and hitch up to the trailer in the normal way. Then we retraced our tracks and went south to Diyarbakir and turned up towards Lake Van. That road eventually brought us out at Agri where we picked up the main road to the border with Iran. It was a hell of a diversion.'

Graham Wainwright still has vivid memories of the harsh winter conditions he and the others regularly endured as they battled through Eastern Turkey. He clearly remembers how he and his mates used to cope with the very basic facilities as they struggled across the wilderness:

'Believe it or not, we used to get black from working outside in the freezing snow. It was so dirty and used to get ground into the skin. We'd stay like that for days if there was nowhere to wash.

'When we finally found a waterhole we'd have to spend a while breaking the thick ice off the top, then jump back into the cab and put the heater and blower on full blast. Then we'd strip naked and jump out of the cab with a load of shampoo already on our hair. We'd throw buckets of ice-cold water all over ourselves and have a very quick wash, then jump back into the cab, shivering, to dry off and warm up.'

In the early days, there were no proper stopping places for the drivers and it was a case of finding a suitable place for a wash and brush-up. Peter Cannon said that the fresh mountain-spring water which ran constantly into troughs or straight out of the pipes in the mountainsides was very invigorating. It was especially nice during the hot, dusty summers but something to avoid in the harsh winters: 'Some of the men were braver than me,' he admitted.

In Eddie Parlane's view, 'In winter the job was certainly

an adventure and we gained some good driving skills going over the mountains. If we got through unscathed, we considered ourselves to be high-class lorry drivers.'

And as Bob Paul concluded:

'We were just trying to do our job and wanted to get through to the other end, but at times it was nigh on impossible. Eastern Turkey was a terrible place. We had some horrendous conditions to fight against in those early days, especially in the winter and springtime, but we all just got on with it. My boys were magnificent in the way they coped with everything – but of course it was all in a day's work for us.'

With modern bypasses and motorways in place, the adventure has gone and the job that the Asian Transport boys were doing will never be repeated. Gordon Pearce had more than his fair share of adventures in Turkey, as recounted in Chapter 12. 'Everyone had places which they feared,' said Pearce, 'but the toughest of the lot for me was Tahir.'

'The summit was over 10,000 ft above sea level. Starting from its base at 6,000 ft, the actual climb was a little over 4,000 ft. Not a lot you may think – but imagine driving a fully loaded 40-ton rig that is about 60 feet long up a very narrow and twisty track to the top of Ben Nevis. That's what I would compare Tahir to.'

Whether the hazards in Turkey were connected with the trucks, the people or the terrain, they were always evident, spring, summer and winter. Nevertheless, the

Left to right: Peter Cannon, Tony Soameson and Bobby Vallas. No truckstops in this part of the world, just a trough full of ice-cold mountain water. (TS)

sense of humour which the men had was something which always kept their spirits up. Eddie Parlane recalled one such humorous occasion:

'I was paired up with Gordon Pearce in UHM 25F and my brother Bun was with Graham Wainwright in UHM 26F but they were driving slightly further ahead of us. The

Bobby Vallas poses with UHM 26F at the top of the formidable Tahir mountain pass in spring 1970. (BV)

John Frost, driving UHM 26F, begins the long climb up Tahir in summer 1969. (JF)

Once the bulldozers had cleared a path, the rigs were able to continue. (GP)

weather was particularly bad and it had been raining very heavily for many days. As we negotiated one of the many mountain passes, we came up against a landslide which had only just happened. The road had been completely covered by earth and some huge rocks, so no one could get any further.

'When we caught up with the other truck there was no sign of Bun or Graham in it. Gordon being Gordon decided to put the kettle on, but I wanted to go and see where they were and how bad it was so I could get an idea of how long we might have to wait.

'It was terrible weather and my trousers were ripped and covered in mud. My overcoat was very dirty and full of holes. I was wearing an old hat which was almost falling to bits and I hadn't had a shave or a wash for days. As you can imagine, I looked awful.

'As I walked over part of the landslide, I spotted Bun and Graham talking to other people. They were all having a big conflab and were very engrossed in deciding what to do, so I casually walked over to Bun who had his back to me and I said in a funny accent, "Cigarette, cigarette?"

'Graham recognised me but didn't give the game away. "Blimey," he said to Bun. "Look at this bloke. He's in a right old state and trying to scrounge fags. Can you smell him, Bun?"

'My beloved brother didn't even turn round. He just said, "Give him a fag to shut him up and send him on his way."

'"Oh, thanks, old mate," I said. "Is that all you think of me then?" It was hilarious and Bun was really embarrassed. I never let him forget it.'

Dick Snow's first trip to Tehran was certainly one to remember. It was December 1968 and he was paired up with Ray Scutts when they got stranded in the

A huge landslide halted everyone's progress. UMH 26F (front) and UHM 25F wait patiently in the queue. (GP)

middle of nowhere in Eastern Turkey for almost ten days. As Scutts explained:

'The pair of us ended up having Gordon Pearce and John Frost looking for us. It was so cold where we were that the engine had frozen and even after we bypassed the fuel filters with a length of garden hosepipe to send the fuel straight to the pump, the old Scania still wouldn't run.'

As the weather began to worsen to blizzard conditions, Scutts chose to take a different route to Pearce and Frost with whom they were running in convoy. Scutts thought he would get through more quickly, but unfortunately he ended up stuck when he got caught in a snow whiteout where he couldn't tell land from sky.

Not wanting to move any further, he and Snowy then decided to leave the truck and walk to civilisation which he thought was about four miles away, back on the original route. Snowy wanted to make a set of skis but as Scutts said, 'What did he think he was going to make them out of? We had nothing.'

Scutts knew there was a PTT telephone exchange in the nearby town. However, as they began to walk they became aware that they were being followed by some wild dogs, so they returned hastily to the safety of their truck. All they had to keep themselves warm and boil water was a tiny gas burner.

Scutts thought it best to stay where they were and wait for help as there was no let-up in the bad weather. As the days passed, the pair ran out of food. They only survived on snow which they melted and boiled, and to which they added Bovril. Scutts remembers vividly that both he and Snowy looked and felt quite ill and had lost some weight while they were waiting to be rescued.

When Pearce and Frost finally came across the pair, both Scutts and Snowy were suffering from hypothermia so a good strong drink was called for. Scutts said he was not a drinker, but:

'I tell you, that whisky was the best thing I'd ever drunk in my life. I think I was pissed after two gulps. Snowy on the other hand had two-thirds of the bottle and never even looked tiddly. In fact, even though he was not feeling too good, he still thought being found was a good enough excuse to strip off and run around in the snow. He was totally mad.

'We were all so bloody pleased to see each other and of course the others had plenty of food so they cooked us a delicious meal of Campbell's meatballs, tinned new potatoes and a packet of Smash. I'll never forget it, it tasted like caviar.'

Ray Scutts, driving WLO 95G, got caught in a whiteout after taking the wrong road. (RS)

Even though he was the boss, Bob Paul was never one for shying away from problems and would always take responsibility where necessary. One example was when he and Pearce were paired up in the AEC and Graham Wainwright was on his own in a hired Leyland tractor unit with trailer. In fact it was Wainwright's very first trip for Asian Transport.

The vehicles were in convoy in Eastern Turkey on the way to Iran, but the Leyland had developed problems with its brake drums which had all but fallen off due to the terrible road surfaces. Pearce and Wainwright had made some running repairs before they started to tackle a long descent down a mountain pass. As it was his first trip, Wainwright felt a little unsure of driving down such a difficult stretch of road with brake problems and was worried that they wouldn't be good enough. Paul told the story:

'As the AEC had less weight in it, I told Pearce and Wainwright to go ahead in that and wait at the bottom of the pass for me while I nursed the Leyland down the five or six miles of hills. I told the pair that if anything were to go wrong then it should be me who dealt with it as I was the boss. I waited a while after they had gone and then set off myself. I drove very slowly to begin with and tested the brakes every few yards.

'Everything seemed okay, so off I went. I knew where they would be, so as I came round the last bend I put my foot down hard, flashed the lights, sounded the horn and waved my arms frantically. As I shot past, all I saw was a flash of Wainwright's boots as he dived out of the way. He obviously thought the brakes had completely failed and I couldn't stop. He would never admit it, but I know I frightened him to death!'

The last section of the journey through Eastern Turkey took the men past the small town of Agri. From there it was about a hundred miles to the Turkish exit border at Gürbulak. Since Ankara, the roads were mainly dirt and quite mountainous, but here for the last fifty miles or so, there were good tarmac roads as it was an important military area. To the south, there were the foothills of Kurdistan and to the north it was

John Frost (left) celebrates with Dick Snow, looking a little thin, who thought it was a good idea to strip off! (RS)

only thirty miles to Georgia and the Russian border. Across the intervening plain, there was a clear view and the bold landmark of the 16,786 ft Mt Ararat which, even in summer, is always snow-capped.

The Turkish exit border at Gürbulak and the Iranian entry border at Bazargan constituted a small but very important complex. Asian Transport employed a Turkish agent there who looked after the trucks, the documentation for the loads and the drivers themselves as they passed from Turkey to Iran and back again. As Graham Wainwright said: 'We all got on very well with the agent, so much so that I used to take him Western goods such as Grundig tape recorders, biros and jeans. He was really good to us while our documents were being processed.'

There was a strict protocol to enter the border complex. A Turkish soldier would allow trucks to pass through a single-width archway into a courtyard which

AMY 147H (left) and UHM 26F. Peter Cannon poses with the awesome Mount Ararat and its little brother in clear view. (TS)

was approximately 120 yards square and divided into two by a low wall topped by 4 ft steel railings. One half of the courtyard was the Turkish area and the other was the Iranian one.

In the middle of the wall was a 12 ft gap where a chain hung and was unhooked whenever a vehicle needed to pass from one side to the other. Two Turkish armed soldiers guarded their side and two Iranian armed soldiers looked after theirs. Around the outside of the courtyard were the buildings of the customs offices, warehouses, sleeping accommodation for the troops, passport control, etc. At the opposite end of the courtyard was another archway, again guarded by soldiers through which trucks headed into Iran.

Peter Cannon remembered the complex well:

'In my experience, it was rare that the Turks wanted to offload anything on the way out of their country, although it did happen from time to time. The Iranians, on the other hand, would check the whole load on every trip as we entered their country.'

Clearing customs to leave Turkey did not take long at all. Gordon Pearce remembers making friends with a Turkish customs officer to whom he used to present a couple of packets of cigarettes plus the odd goody which the officer had asked Peace to bring him from England:

'I never thought of it as bribery and corruption, just being friendly. It seemed to speed things up, though. Once the documentation was cleared and stamped and my gate pass shown to the soldiers, the chain was lowered and I was directed to the Iranian side.

'There were six wooden ramps to get the rigs level with the customs warehouse floor. We had to reverse the trailer onto one of them, unhitch it and then reverse the truck onto another which was a tricky manoeuvre on a dry day and even more awkward when it was raining or icy.

'The Iranians would then proceed to unload everything into the warehouse and go through it all with a fine-tooth comb, checking it all against the paperwork before putting it all back into the truck and trailer again ... well most of it, anyway.

'When we drove into the Iranian side of the complex we would regularly sit for two or three days waiting for our loads to be checked and for customs clearance before we could continue to Tehran. The only good thing about the place was the restaurant where we could have a good meal and a few drinks.'

A view from the Iranian side into the Turkish area showing trucks backed up to the warehouse (right). An armed Iranian soldier and the dividing chain are just visible. Note the Hungarocamion rig entering from the archway. (BV)

The TIR Carnet system of guaranteeing goods which was used throughout Europe finished at Gürbulak. Once the trucks entered Iran, a different type of guarantee was implemented – hence the laboriously long checking process. The Iranians used a cash guarantee and it could sometimes take up to a week to arrange.

The load manifest also had to be translated into Persian so all the drivers could do was sit and wait. The reason for the long delay was due to the extremely high duties that the Iranian government put on imported freight. Some of the loads were very valuable so the authorities wanted to be absolutely sure that what was in the trucks was exactly what was written on the paperwork.

Johnny Walker Whisky, for example, was a major customer for Asian Transport in the early years. The trucks were taking 40 tons of the expensive liquid to Tehran every month so the authorities wanted to be sure there was no smuggling into their country.

Having spent a couple of days at the border being checked, the men would then be free to drive out of the compound. However, they would have to wait again on the road outside for the agent in Tehran to provide a bond to the value of the cargo on the truck before it was allowed to move. As Cannon explained:

'If another one of our trucks was up ahead we had to wait until it had entered the customs compound in Tehran and discharged the bond so that the next truck could use it. Sometimes we were stuck at Bazargan for a few days if it was a weekend or a public holiday.'

While they waited, the trucks were parked on the road in a haphazard way and there were always a lot of them waiting, often thirty or forty. 'It gave us a chance to meet other drivers,' Pearce said:

'We got on well with the Iranians and Bulgarians. Quite a few Armenians worked for the Iranian transport company TBT and the drivers spoke up to six languages – Russian, Persian, Turkish, German, French and English. We could understand each other by using words we knew in a bit of one language or another. It seemed to work both ways but I suspect it confused the locals.

'The TBT drivers were a very friendly bunch and we got to know some by name, most being called Uri, and that made meeting up on the road a real occasion. We'd all park up together and enjoy a communal meal. Each of us would

Peter Cannon, sitting by YYW 330G, rests with friendly Bulgarians and Iranians at the restaurant in Bazargan. (PC)

prepare a different course and we'd give each other little gifts from our particular countries. That was what made the whole job so special. It was the camaraderie and friendliness of everyone which made it unique in the early days.

'The restaurant at Bazargan was lovely. We'd sit and talk to the other drivers until the early hours over a meal and a good few glasses of raki.

'Then it was good to get away from Bazargan as we knew it was only another five hundred miles to Tehran. On one trip I got cleared extra quickly because my agent's wife was expecting a baby and he desperately wanted to be with her. So, with just three quick coffee stops and a few biscuits to keep me going, I did the journey overnight in eleven hours. My Scania Vabis went like a dream at a maximum of 56 mph and I had the perfect, uninterrupted run.'

When the men finally arrived in Tehran, they took their rigs straight to the customs compound where their cargoes would be cleared and unloaded. While the men waited, they would usually stay at the Hotel Continental for some rest and relaxation. Cannon described what happened next:

'The rigs were always offloaded in the customs compound and we never had to deliver anything anywhere else. For our return loads we would pick up at various warehouses in and around Tehran which made things nice and simple. Expensive Persian carpets were our main loads which we picked up right opposite the customs compound.

'Gum tragacanth, pistachio nuts, some personal effects and other bits and pieces would form the remainder of our cargoes. We would be given a huge load manifest by the agent and would then spend a long time in the hotel writing out the TIR carnets by hand for the return journey. There were about eighteen copies to make, but with the sheets of carbon paper we used we could only press through about three pages at a time. So we had to do it several times as every page had to be legible. With a long list of different goods it often took quite some time.

'On the way home, we would try and leave Tehran very early in the morning, as it would take twelve or thirteen hours to get back to the border at Bazargan. Exiting Iran was no problem, but we would try to arrive "over the chain" into Turkey before the customs office closed at 5 pm. Luckily we had to put the clocks back one and a half hours at the border, so we gained a bit of extra time there.

'The TIR carnet started as we entered Turkey at Gürbulak. If there were no problems with the documentation, we could do another two hundred miles to Erzurum that night, giving us a good start on the homeward journey. That was of course only possible during good weather, as we would certainly not attempt some of the mountains at night during the winter, especially in the snow and ice. Well, at least I wouldn't.

'I would dread the winter trips coming up, but when you eased yourself into them, they actually didn't seem too bad. Like everyone else, I was certainly well organised on my trips. Somehow I didn't experience the amount of trouble that some of the others had, although there were always breakdowns and lots of punctures to contend with. The solutions to some of the snow and ice problems were in your own hands and you needed to know when to chain up and when to park up.

'You also needed to help yourself. You had to see to all the little things like having a supply of alcohol for the brake system in your toolbox, keeping a supply of inner tubes and valves close to hand, a spare gas bottle, and the all important tools. I would spend some of my time off in the UK organising these things, whereas some of the others would just go on the piss.

'We all used to get depressed about the job at times, especially during the harsh winter months. I used to think, "What am I doing here when I could have a nice cushy UK job and probably earn more money?"

JLL 686K, driven by John Williams, is about to take on a load of expensive Persian carpets in Tehran. (ASTC)

'But after I'd got over the problem, whatever it was, I'd get a great feeling of satisfaction. My confidence would return and then I'd be ready for the next challenge which was what the job was all about. Most times I even looked forward to it.'

Loading the carpets was very hard work. (GP)

Everything was hand balled. Note the bale weighs 152 kg; some weighed more at 210 kg! (GP)

Tony Soameson, driving AMY 147H, is having fun racing a double-engined steam freight train in Turkey. (PC)

AMY 147H and UHM 26F are resting while Tony Soameson tries his hand at some skiing. (PC)

Party time: Graham Wainwright and Ray Neale. sharing WLO 95G (left), meet Peter Cannon and Tony Soameson, sharing AMY 147H, going the other way in Eastern Turkey. (GW)

AMY 147H suffers a front blowout in Turkey en route to Pakistan. (PC)

John Williams, watched by a local Turk, tries to get the engine of JLL 686K started after the fuel has waxed up. Note the small fire under the cab. (DP)

Gordon Pearce, driving UHM 26F fully chained up, follows the river between Erzincan and Erzurum, Eastern Turkey. (GP)

Graham Wainwright (photographer) and Bun Parlane (wearing cap) stopped to assist an Astran Swiss subbie, Rudi Rusterholt, who'd been involved in a crash with a Tonka. (GW)

CHAPTER 12

Gordon's Adventures

On 13 December 1966 Gordon Pearce set off to Tehran, the capital of Iran, on his very first trip to the Middle East. He and Bob Paul were sharing the new AEC drawbar outfit on its inaugural run. 'We drove to Tilbury for a ferry,' said Pearce, 'and our families were there to see us off. Michael Woodman came too and it seemed like a real occasion.'

In that era, dock rules stated that all vehicle movements on and off the ferries should be done by the dock workers, but when they saw the huge Asian Transport rig pull up, they suddenly changed their minds and told Pearce to put it on the ferry himself. After a couple of manoeuvres and some careful steering to line himself up with the link span, Pearce reversed the rig perfectly into position inside the cavernous ship. The ship's crew even clapped in appreciation. Pearce continued:

'Everyone waved as we headed out towards Rotterdam. I felt like royalty. It was a little strange, though, as all we had with us for such a long journey was a suitcase of clothes each, some basic food supplies and a few spares and some tools.

'I admit I was quite nervous because I didn't know what lay ahead, but I put the nerves out of my mind because Bob had been out previously and knew what he was doing. Anyway, I was thoroughly enjoying every minute of my first adventure in my new job.'

The Asian Transport rig was only lightly loaded with about six tons of vehicle and electrical parts, baby food, cosmetics and some personal effects. As Pearce and Paul were both competent drivers, they were able to travel at the rig's maximum speed of 40 mph on the flat and, if they were lucky, up to 50 mph downhill – out of

The AEC performed well on its inaugural journey. Here it is at rest in Austria as Bob Paul empties a water container. (GP)

gear! As they trundled across mainland Europe, both men settled into driving but kept swapping duties so each of them could get plenty of rest.

The new AEC rig was performing well, and all went smoothly until the men stopped for a break near Belgrade in Yugoslavia:

Some of the roads were bumpy, to say the least, so we stopped to check the truck over and when we got to the toolbox it was open and everything was gone. The spares and my tools had all fallen out. Someone had left the lock undone and it had fallen off. We had lost important things like the jack, a fan belt, hoses and bulbs and we couldn't do without any of it.

'There was nothing else to do but retrace our journey and see if we could spot any of it along the roadside. We turned around and drove slowly back for twenty miles or so looking carefully for our lost equipment. We did manage to find a few bits but also got ourselves into a couple of nasty arguments with the locals who had come across our stuff. It was very difficult trying to convince these people that the bits were actually ours, so we gave up the cause. We decided the best thing to do was try and re-stock as we went along.'

The pair headed through Eastern Europe and into Turkey where their problems really started:

'As we entered Turkey the roads were reasonably good to Istanbul and then got gradually worse towards Ankara. The Turks were always re-building their roads but unlike us with our big yellow diversion signs, they used stones to indicate a "detour" and it didn't matter if they were big or small stones, they were simply scattered across the road to indicate that it was closed.

'We came across one such road closure but before either of us realised what it was, we had driven straight over a row of small stones.'

Pearce continued driving, but the road then stopped abruptly where a bridge was being constructed. In front of him was a twenty-foot drop straight into the river bed:

'I had no choice but to reverse all the way out. It must have been at least two or three miles and took me a good couple of hours. It was hard work with no power steering. I had almost to stand up to pull the wheel round. It certainly tested my reversing skills.

'While I was trying to keep it in a straight line, I recalled what the expert at Dereham had told me about reversing the rig. He said the only way was to push the trailer backwards with the nose of the truck. I wished he'd have been with us because I wondered how he would have turned the truck around with no room to move either side.

'Bob never did get the hang of reversing that well, so after that episode he always left the reversing to me as he said I was the expert.'

Pearce and Paul continued on past Ankara where the weather began to get extremely cold. Pearce recalled his job interview:

'Ah yes, the hot, sunny Middle East. Except it was freezing by now at minus 15 centigrade and getting colder by the minute. I couldn't believe it.

'Then I suddenly realised the rig was slipping and losing traction. We were on an uphill section of road where hard-packed snow and ice had formed and made the surface incredibly slippery. No matter how hard I tried to keep it going, the rig finally came to a standstill. Then I felt it start to slide backwards....'

As the tyres were warm from driving, as soon as the rig stopped they melted the crust on top of the hard ice which caused the rig to start sliding backwards on the smooth ice underneath. Pearce again remembered what the AEC engineer had told him about diff locks and Michelin tyres:

'We engaged the diff locks but it made no difference at all, the wheels just kept turning. The truck wasn't going anywhere on the ice so the engineer's words were futile.

'I had brought about eighteen feet of common toilet chain which I thought might help in such conditions. We wrapped it round the drive-axle tyres as best we could, but alas it didn't last five minutes, it was useless.

'We had to scrape the snow and ice away from the side of the road to get to the dirt beneath which we put under the wheels and then, very slowly and carefully, we managed to get ourselves going again. It took hours.'

When they stopped at Kaman, a village south of Ankara, Bob Paul decided to splash out and buy a set of Turkish heavy-duty snow chains.

Although the men were travelling during the winter, daytime driving was relatively easy because there were beautiful blue skies and a fair amount of warmth from the sun. However, the days were short and as soon as darkness fell in the afternoons, the temperature dropped drastically. The truck became sluggish, losing power.

Pearce recalled what the engineer had told him about diff locks and Michelin tyres. In reality, the rig was going nowhere! (GP)

Pearce had never driven in those conditions before and thought it must have been some sort of fuel starvation or, worse still, he wondered if the truck was running out. There was no fuel gauge on the dashboard, so he got out and checked the gauge built into the side of the tank. It showed all was well but when he shone his torch inside to check, he noticed the fuel looked waxy.

'I hadn't heard of the waxing up in which diesel fuel becomes thick and unable to pass through the filters,' Pearce admitted. 'And I didn't know about lighting fires under the fuel tanks to warm the fuel and thin it out again.'

Pearce and Paul kept driving as far as they could until the truck finally stalled and wouldn't re-start. They had no choice but to spend the night right where they were, perched on a narrow road half-way up a mountain pass.

Pearce lit the Calor gas stove and warmed up some tins of soup which he said tasted like a gourmet meal as both men were very cold and hungry. Steaming mugs of tea followed and, after discussing the day's travelling, the men wrapped themselves in blankets and sleeping bags and tried to get some sleep.

More through luck than judgment, they had parked facing north, so as the sun came up in the east the following morning it shone directly on the metal fuel tank. A couple of hours later the diesel was warm and thin enough to flow through the pipes which allowed the engine to be started.

Note the marks on the ice where the rig had slid backwards. The trailer began to jack-knife too. (GP)

From then on it was really only the elements that caused the truck problems and on Christmas Day 1966, Pearce and Paul arrived at the border out of Turkey. After processing their documents and passports at Gürbulak, they drove the short distance through no-man's-land and repeated the procedure at Bazargan to enter Iran.

When Pearce was packing his supplies into the AEC

As temperatures dropped, the fuel became waxy and the rig eventually stalled. Pearce and Paul spent the night on this narrow mountain track. (GP)

before the trip started, Paul asked him what was in a particular box, to which Pearce told him it was personal. While they sat at the Iranian border on that Christmas morning waiting for clearance to drive on, Paul was taking some sleep.

Pearce took this opportunity to open his box. It was full of tinsel and streamers with which he decorated the whole cab. Paul was delighted when he woke up:

'It was absolutely wonderful. I opened my eyes to see all the decorations and if that wasn't enough, Gordon had brought a tape machine which was playing "Jingle Bells". But to cap it all, he'd even packed a Christmas pudding complete with custard. The Iranians couldn't understand what was going on. They thought it was someone's birthday.'

After the celebrations the pair got permission to drive on and it took just another couple of days to roll into Tehran. The men found the customs compound in the city and parked the truck there while the documents were processed and everything was unloaded straight into the warehouse.

Pearce and Paul were not allowed to stay with the AEC while it was in the compound, so they set about finding a local hotel where they could have a good clean-up and relax in a comfortable bed. That wasn't easy, as Pearce explained:

'On the first trip we ended up in a cheap, run-down place which had no hot running water. Then on the second trip we found a slightly better hotel but that had a noisy club underneath. On the third trip we came across the Hotel Continental and that was it. We had found the perfect place

Christmas Day 1966. Pearce decorated the cab and even had a Christmas pudding and a tape machine playing carols, much to Bob Paul's delight. (GP)

to stay and always used it from then on. We liked it best of all because the English dancing showgirls all stayed there too.'

The first journey from Tilbury to Tehran had taken just fourteen days during which the men had covered almost four thousand miles. After the rig had been unloaded it was taken to the British Leyland garage nearby to be serviced and repaired. Relieved to have got there in one piece, Pearce thought he could at last spend some time relaxing and exploring the vibrant new city, but unfortunately for him, Bob Paul had other ideas:

'While Bob and I waited for the truck, we donned our suits which had been folded neatly all the way from home and set off to try and find a return load. We used local taxis as they were extremely cheap and the drivers knew where all the freight agents and manufacturers were.

'After we'd knocked on quite a few doors, we managed to secure a load back and only then did Bob let me off to explore the markets and bazaars which I found fascinating. All the traders were inviting me to take cups of tea with them while they plied me with their wares. Even though they spoke in a strange language, it was great fun and I remember bartering with one old fellow which I really enjoyed.

'I ended up with a really nice rug but I hadn't got enough money so he let me take it away on the understanding that I returned next time with the cash. He didn't know me, but he knew I'd come a very long way and he trusted me which I thought was incredible.'

Trade relations between the UK and Iran were very good at the time. The country was under the rule of the Shah and, as Pearce put it: 'We felt as safe as houses.'

After a couple of weeks in Tehran, the rig was finally re-loaded with expensive Persian carpets. Pearce and Paul left the city in high spirits and began the long journey home. As they entered Turkey, the weather was just as it was on the outward journey with beautiful clear blue skies during the daytime:

'We were travelling at a height of about 6,000 ft and we had only driven twenty to thirty miles into Turkey. Suddenly, as darkness began to fall, the temperatures plummeted, but this time to a mind-numbing minus 30 degrees centigrade.

'I thought Bob was experienced with the cold weather from his trips to Afghanistan, but in fact he knew as much as me and that wasn't much at all. He had done his trips during the spring and summer and had never come across such cold weather.

'Neither of us knew much about heating the diesel to keep it flowing from the tank but we now understood that the engine was losing power because the diesel was getting thick and clogging the filters up. Through trial and error, we found that by keeping the revs down, we could just about keep going. The trouble was, we ended up driving on tick-over and had no power at all.

'We tried to get as far as we could that evening, hoping to find a safe place to stop for the night. We struggled on with howling winds coming in through the gaps in the doors and around the pedals. Then, as we climbed a hill, the truck conked out, just as it had done on the outward journey.'

Again the men found themselves stranded on the edge of a narrow track, totally isolated and getting colder and colder by the minute:

'We decided to try and have something hot to eat. I managed to get the stove lit and heated up some soup but unfortunately it got knocked over onto our blankets so we only had a small amount each.'

As the clock ticked on and the temperature fell even further, Pearce realised that he had to take evasive action and preserve what precious commodities they had. He wrapped up as best he could and climbed out of the cab. Braving the harsh elements, he lay under the truck to try and drain the water and antifreeze from the engine. He managed to catch most of it in a bowl and a couple of buckets, but it wasn't long before that started to freeze.

No other traffic had passed since they had become stranded so they resigned themselves to spending the night where they were. They sat fully clothed on the cramped bunk, huddled closely against each other and wrapped in blankets to try and keep as warm as possible. According to Pearce's magnetic thermometer perched on the cab roof, it was now minus 35 degrees centigrade. It was impossible to relax or sleep.

When daylight came, the men saw that the skies were cloudy and misty with freezing fog. There seemed very little chance that the sun would appear to warm things up. Pearce wrote the following stark entry in his diary: 'All hope gone'!

'Even so, I wasn't really worried. We knew a local truck would come along eventually because it was the only road in the area. Anyway, there were two of us to keep each other's spirits up.'

While they sat huddled together in the freezing cold

In Eastern Turkey on the return leg from Iran in January 1967, the AEC is fitted with a full set of snow chains. By now, Pearce and Paul were accustomed to the harsh winter conditions. (GP)

cab, one solitary truck did go past, but as it was going up the hill, it didn't, or couldn't, stop to help.

Later that day and much to the relief of Paul and Pearce, they spotted some men approaching with a bulldozer. 'Maybe the truck that passed us earlier told them where we were,' Pearce suggested.

The Turkish rescuers were based at a grader station not far from the stranded men. They chained the complete rig to their machine and towed it up the hill to their compound which was only about half a mile further on.

The English drivers could see the top of the hill from their cab and it would have only taken a ten-minute walk to have reached it - but of course they had no idea what was ahead of them. If only they'd known!

'We stayed with the workers in their hut, sitting by the lovely warm fire and then had a good night's sleep in their spare beds. They really looked after us and next morning they started to free up our engine.'

The drained fluid now had to be replaced before the engine could be started. Unfortunately most of it was spilled, so Pearce and Paul took a jerry can and walked to a nearby river. There was a foot-thick covering of ice on it but the local villagers had bored a small hole through which Pearce was able to fill the can:

'As we trudged back to the hut, I became breathless and began to struggle in the deep snow. It occurred to me that it was because we were at a very high altitude so Bob and I swung the can to and fro to get momentum to help us walk, then it suddenly started making clunking noises. I looked inside to find the water was turning into ice as we walked along.

'Then as I breathed out, the moisture from my breath was freezing on my beard and moustache. When we got back to the hut we had to leave the can in front of the fire and sit there ourselves for a while to thaw out.

'Eventually, we were able to mix the antifreeze which I'd saved with the river water and poured it back into the engine. The Turks helped us by putting some methylated spirits in the fuel tank to help break down the waxy diesel.'

The Turks lit a huge fire under the truck's fuel tank which amazed Pearce as he thought it would explode.

Disembarking from the Bosphorus ferry on the western side of Istanbul. From now on, the journey home would be a piece of cake. (GP)

It was left to burn while the Turks used a small blowtorch to gently warm the fuel pipes. One of them jumped in the cab and tried the ignition a few times. Eventually the engine started and after a few coughs and splutters it began to idle properly. Pearce continues:

'As he was already in the cab, the Turk managed to find a gear and move the truck forward. We shouted at him not to stop as the fire was now under the back wheels and one of the tyres caught alight! We shouted at him to keep going forward but of course he didn't understand.

'There was a lot of shouting and arm-waving as we all tried to get him to move. Finally he realised and managed to get the truck away from the fire after which we all began throwing snow on the burning tyre to put it out. We saw the funny side of it and luckily there was no real damage.'

Pearce and Paul were guests of the grader station men for a couple of days and realized that without the men's help they might well have ended up with hypothermia. 'I owed those workers a great deal,' Pearce said. After farewells and handshakes all round, Pearce and Paul were able to set off on their homeward journey.

After their new experience with the severe Turkish weather, they were now prepared for anything as they travelled westward towards Istanbul:

'We felt much more confident to deal with things. Even though we got caught out on the snow and ice a few more times and jack-knifed the rig on more than one occasion, we were able to cope and get ourselves back on track.'

After this the men enjoyed an uneventful trip home, completing the journey to England by ferry from Rotterdam to Felixstowe where Michael Woodman was waiting to greet them.

'It was a huge learning curve,' said Pearce:

'I learnt an awful lot about the weather conditions which I was utterly amazed by. I really was expecting the Middle East to be hot and sunny. I never realised it could get so cold, even in winter. I was pleased to learn from the locals how to cope in the freezing cold and what to do with the rig when we froze up because I knew I would need that knowledge on future trips.'

For about eighteen months Pearce accompanied Bob Paul on trips to Iran. Although he has many memories of those trips, there are some which stick more clearly than others in his mind:

'There was a stretch of road before we got to Tehran that was dead straight for about eighty miles. However, it was incredibly bumpy and corrugated, so we could only drive for

about an hour and a half before we had to swap over. We had to concentrate continuously, always looking for the best bits of road to drive on.

'We drove about twenty-one miles each at a top speed of just 10-15 mph. It was too much for the suspension and steering to cope with if we went any faster than that, so it took about eight hours to get from one end to the other.

'We did about six trips over that same stretch of wretched road, then one day we found that they had asphalted it all. It was marvellous. We did the eighty miles in an hour and a half and immediately saved about six hours on the journey.'

Pearce enjoyed driving the AEC immensely:

'It was a fine old workhorse and served us really well. We got quite attached to it and knew exactly how to drive it. We knew how the engine should have sounded at any given speed.

'I only once saw Bob Paul become frightened. I was driving on a mountain road while Bob was sleeping on the bunk when I had to negotiate an extremely tight hairpin turn with a nasty camber. It really did need a few shunts to get the drawbar outfit round.

'Poor old Bob was totally exhausted and was obviously in a deep sleep, but the jolting and the rocking of the truck and the stark difference in engine noise as I was constantly revving up then braking must have disturbed him.

'I had to put the front of the cab extremely close to the cliff face in order to give the trailer enough room to follow round and at exactly the point that I was half-standing and pulling hard round on the steering wheel, Bob woke up with a start.

'All I heard was a very loud scream directly in my ear. As soon as he'd opened his eyes he must have seen the rock face dead ahead and presumed we were going to crash into it. The fact that I was half-standing and trying my hardest to get the rig round the corner made matters worse for Bob. He jumped a mile and told me later he thought we were going to die.'

Paul does remember the incident clearly – in fact he still has occasional nightmares about it!

One of the regular problems from which the AEC suffered was that the injector pipes used to break away from the collars that they sat in when the truck traversed the rough roads. The problem seriously impaired the power and speed of the truck. Over the months that Pearce drove it he became adept at fixing the pipes.

However, on one occasion he and Paul were coming through Holland without having had chance to stop and make the necessary repairs:

'We found that if we kept the revs down we could keep going and so we were chugging along at around 20 mph when we got stopped by the police. They wanted to know why we were going so slowly on their roads which were designed for fast traffic.

'They understood the problem and said as long as we could travel at 64 kilometres an hour (40 mph) we could continue. Bob and I looked at each other and tried not to laugh because we knew the truck wouldn't do that even flat-out, but we didn't want to tell them.'

After his first trip with Bob Paul in 1966, Pearce soon found his vocation and was doing something that he really enjoyed for the first time since leaving school. He took to driving the AEC drawbar outfit like a duck to water as he continued with his trips to and from Iran and then, from 1970, his trips to Doha.

He made many friends along the way, especially with the officials at the borders and the agents in the towns who were arranging the back loads. Getting to know those people saved him time and money on future journeys:

'When I came back home I used to spend a whole day shopping for people. I bought tomato sauce for a man at the British embassy in Ankara, ink for someone for whom I'd previously bought a pen and some paper. There was Milk of Magnesia for a waiter in Bulgaria and fishing line for a Yugoslavian man who couldn't get the right size.

'I even got some old used stamps for an Iranian boy who lived in an earthquake-torn village. I used to take regular supplies of lighter fuel to a customs official to whom I'd given the lighter a few trips beforehand. It was never-ending.'

On the outward journeys entering Bulgaria from Yugoslavia, there was a duty-free shop which only foreigners could use. Pearce and his colleagues would buy a litre of 10-year-old Ballantine's whisky for around two pounds:

'It was beautiful stuff. We were allowed one bottle each which we used for bribery if we were desperate at somewhere like a lengthy border hold-up, but if all went well and we didn't need to use it, we'd drink it ourselves on the way home.

'I remember bumping into Dick Snow and Bobby Vallas coming the other way once so I stopped to see them which I knew would be a fatal mistake. That was it for the rest of the day. We got the bottle out and had a good piss-up. It was rare we met each other like that so we made the most of it.'

Gordon Pearce thoroughly enjoyed his adventures, especially driving the AEC. (GP)

Being quick-witted was the attribute drivers needed to get through some of their predicaments. Pearce became a master at this and he got himself out of many sticky situations with a mixture of bluff and skill. Here is just one example.

At some of the border crossings the customs office had a series of windows behind which an officious stony-faced man sat, waiting for various documents to be passed to him. He then scrutinised, stamped and threw them back at the driver. The windows had to be approached in a certain order so as to get the documents stamped correctly.

On one trip at the German/Austrian border Pearce discovered that he was missing one important document just as he approached the first window. Quick as a flash he chose to start the process at the middle window, therefore missing out the first part of the sequence. Once he'd done that, he stood back and quietly waited a while for the shift to change.

Once he was satisfied that the original men had gone, he went back to the first window and blamed the previous shift for losing his document. With a smile and his polite manner, Pearce made his excuses to the official who then spouted a whole stream of unrecognisable words at him!

When he'd finished, Pearce calmly said thank you to which the official simply waved him through. 'It was obvious that he was blaming his colleague but I thought it was a case of playing them at their own game and bluffing them,' Pearce said.

A regular problem was the number of punctures. Everyone had their fair share and had to cope:

'I reckon I could repair them with my eyes shut, I had so many. I think my record was sixteen on one trip.

'I became quite an expert at fixing them and once I got the routine sorted, it used to take about hour for each one. In a funny sort of way, having to stop for a puncture was almost a welcome break from the driving.

'The first puncture Bob and I had wasn't at all easy to repair because neither of us had done it before on the AEC. We took the punctured wheel off and then undid two retaining nuts which held the spare wheel up under the chassis. It was fixed on a cable which we had to wind down to lower the wheel.

'Then we put the punctured wheel back on the cable and

John Frost trying to empty the air from the inner tube, near Agri, Eastern Turkey, on the return leg from Iran. The temperature was minus 10. (GP)

wound it up. It was very labour-intensive because we had to get it off again at some point to have it repaired. In the wintertime it certainly wasn't much fun lying on my back in freezing snow at minus15 trying to undo the nuts with no feeling in my fingers.

'On the first trip we only had a bare minimum of tools, but soon bought a set of tyre levers, a tin of 'Tip-Top' patches and started to mend our own punctures as we went along. Although this took time, it meant that as the tyre and wheel were on the floor we could put them straight back on again. Also, we didn't have to start looking round for a tyre-repair garage. Working with a mate made things a lot easier, too.

'All the tyres had inner tubes in those days as there were no such things as tubeless tyres and each wheel had a 'split rim' system of metal rings which held the tyre in place. We would jack the axle up, take off ten wheel nuts – easier said than done – and lay the complete wheel down, split rim upwards.

'Next, we took out the valve to make sure there was no air left in the tyre. Then we used a heavy club hammer to bash hell out of the rim to split it from the wheel. Sometimes we would use some diesel fuel to lubricate the tyre to help get it off the wheel.

'We took the inner garter out, then the tube itself. Usually there would be a nail or a bolt, left by local traffic, which had caused the puncture. We cleaned the area with sandpaper and smeared some glue on it, then fitted the Tip-Top patch, just like a bicycle tyre repair.

'We left it to dry for a while, then fitted the whole lot back together again, making sure not to pinch the inner tube and cause another puncture. We had airlines in the truck which we attached to the engine compressor to re-inflate the tyres.

'With wheel back on and jack and tools packed away, it would be time to make some coffee and warm ourselves up while we made sure the repair held.'

Gordon Pearce worked for Asian Transport/Astran for a little over seven enjoyable years. Although he was meticulous in his ways and extremely careful with everything he did, there were a few occasions when he had no control over his truck, as described in Chapter 14. Nevertheless, Pearce was proud of his achievements. 'My adventures were endless,' he said. 'There was always something new to see and do on each trip.'

He particularly remembered the mountain pass at

Bolu in Western Turkey and being paired up with Bun Parlane in the AEC for a trip to Tehran in the late 1960s:

'In the early days things were very sedate. The trucks were slow and because the pass was such a long drag up, we never went faster than 5-10 mph so we never got out of second gear. It was about six miles to the top and took about an hour and twenty minutes to get there, so we used to have a game of chess as we went along. We carried a set and used it when we were stuck at borders and at customs but it was just as good in the truck.

'We set the board up on the engine hump in the centre of the cab and because we were going so slowly the pieces never moved. I used to wedge a length of wood on the accelerator. The climb was quite straight, so apart from a bit of steering now and again, we could concentrate on the game. We knew almost to the minute how long it would take to get up the hill so we set a watch on the dashboard to time each game.

'We tried a game on the Tahir pass as well, but there were more sharp bends and the road wasn't so good, so the pieces kept rattling and falling off.'

When Pearce was given a new Scania Vabis, he was able to do the six-mile journey in about twenty minutes due to the new truck's bigger and more powerful engine. So it was no more chess from then on.

Much further to the east near Tunceli was another difficult pass close to Pülümür which Graham Wainwright described briefly in Chapter 11. Pearce has vivid memories of the area:

'I was accompanied by Bun Parlane again and we were returning home from Tehran with a full load of Persian carpets in the AEC rig. We were very pleased because we had made good time to Erzurum and were hoping to make the BP Mocamp at Istanbul for Christmas Day – but the weather had other ideas.

'On 22 December 1968, we were on the Erzurum to Erzincan road and as we turned left for Tunceli and Elazig,

Gordon Pearce is bashing hell out of the split rim with John Frost helping, an airline at the ready. (PK)

just before Tanyeri, it was sleeting heavily. As I had seen this weather before I was confident to drive on. It was my ninth trip and I knew the road well.

'The distance between the junction and Tunceli was only about forty miles, but we had an 8,000 ft mountain pass to negotiate before we got there. The sleet began to get worse, turning quickly into thick snowflakes. Despite the worsening weather, we never thought of stopping and waiting for it to improve. We always had the notion to push on regardless.

'The first part of the mountain was a gentle upward slope through the valley following a small river. The heavy snow continued relentlessly, settling quickly. Before long it was about six inches deep and I found it impossible to drive as we were getting a lot of wheel-spin and, apart from that, my vision ahead was almost nil. We had no choice but to stop on an incline. I tried a few times to get going again without snow chains, rolling gently back and having another stab at the hill but it was no use.

'A Turkish coach was following us and he too was struggling. Thirty Turks got out to see if they could help. There were thirty different ideas being discussed and sixty arms all pointing and waving in different directions, so we left them to it.

Struggling to move in the thick snow. Note the tyre marks where Pearce has tried several times to keep going. A Turkish bus behind is in the same predicament. (GP)

'We put our snow chains on the first drive axle of the AEC. By now the snowing had all but stopped so as Bun held the diff lock switch down, I cautiously pulled away and we began to make some very slow progress.

'The next difficulties we encountered were three hairpin bends at the start of the mountain climb proper. The steepness and camber on the first bend was too much for the rig to handle and the wheel-spin soon began to damage the snow chains. I reversed back and had another run in first gear but that was no good. At full revs we were only doing three miles an hour which did not give much momentum.

'We had too much weight on board to even consider changing up to second. The AEC weighed 18 tons loaded and the trailer about 15 tons. As the engine was only 205 bhp, it wasn't very much power at all to pull the whole lot through the snow.

'After half a dozen attempts, we were not even half-way round the first hairpin so the pair of us got out and began digging. We cleared nearly a foot of snow from around each of the six axles and with three hours hard work under our belts, we were still not getting anywhere and the snow had started to come down again.

'We were working against the clock. I decided there was no alternative but to reverse the rig a hundred yards to a wider piece of road where I could leave the trailer in relative safety and try again with just the truck.

'We were totally isolated, there were no villages for miles and no other traffic had been seen for hours. The nearest civilisation was a grader station which I knew was at the top of the pass, about four miles further up the mountain. That was where we decided to head for until the weather improved.

'Leaving the trailer behind, we were finally able to get round the hairpins. By now, the wind had become much stronger and began blowing deep snowdrifts across the road. It was also getting much colder and as we climbed we found ourselves in low cloud. Every fifty yards there were marker posts at each side of the road, but with visibility so poor we wondered which side they were.

'We could only see ten or twenty yards ahead and we were now struggling to spot the next marker. There were no tyre tracks to follow and the wipers had become ineffective against the thick snow which had accumulated on the screen. Although we could now use second gear, even 6 mph

Pearce and Parlane spent three long hours trying to clear the thick snow from around the rig. Parlane can just be seen by the side of the truck, digging while it continues to snow. . (GP)

was too fast in this total whiteout! Would we ever reach the summit? Would we even see it?

'We had to shout over the engine noise to hear each other and we agreed that we had no idea where the next marker post was. I carefully steered the truck in what I thought was a forward direction. Despite the bitter cold, I opened the side window to be greeted by a face full of stinging ice crystals. I strained my eyes and luckily spotted the next post.

'I shouted to Bun that I'd found one and asked him to look out of his side for another. To open both side windows at the same time would let out what precious little warmth we had in the cab, so we did them one at a time. The wipers were now almost inoperable and the screen was covered in snow and ice. We were driving blindly on and, to make matters even worse, it was getting dark.

'Then another problem arose. The engine noise changed dramatically as it started to labour really badly and I found myself struggling to keep the truck moving. I prayed that it wasn't starting to freeze up. I shouted to Bun to close his window so I could have another look out of my side. I stuck my head out and could just make out a deep bank of snow ahead so I promptly stopped the rig.

'It appeared that I had been pushing the snow. No wonder the engine was straining! With a lull in the cloud and wind, I then spotted a steep drop down into the valley. I had been trying to drive over the edge. Thank God I stopped when I did!

'I explained to Bun that there was no road at all on my side so assumed that it must have been on his side. I reversed about twenty feet and moved forward again, steering the truck blindly to the left to avoid the drop.

'Once I had negotiated the bend, we could both see the next marker post, so I aimed straight for it. It was getting quite dark now, but the headlights gave us slightly better vision and with the last hairpin finally behind us, we reached the safety of the grader station at the summit of the pass.

'We stayed with the council workers in their accommodation for a couple of days. We were safe and comfortable and certainly had no intention of going any further until the weather had improved. We were given a bunk bed each and shared meals which were mostly meat stews and vegetables. Breakfast was bread and jam and there was always tea on the go. The dozen workers were very friendly and we really appreciated their hospitality.

The pair finally reached the safety of the grader station near Pülümür. Note the remains of newspaper on the grille, put there to try and stop the bitter cold wind getting into the cab. (GP)

'After a couple of days the storm had blown over so they got the graders and bulldozer out of the sheds and started to clear the road. I went down on a bulldozer to collect the trailer which luckily was untouched. Letting all the air out of the tanks released the brakes and after an hour or so we had the trailer towed up to the top of the pass.

'That was the easy bit. The truck batteries were flat and when we tried to tow-start it, we found that the drive-axle brakes were frozen solid. I cursed myself for having put the ratchet handbrake on when we stopped.

'We set alight some diesel-soaked rag on a stick and warmed the brake drums up. It was just enough to free up three of the four rear wheels. The last wouldn't budge but after a few short sharp tugs from the bulldozer, the engine fired up and the fourth wheel then freed itself after a few shunts back and forth.

'Bun and I hooked up the trailer and made everything ready for the off. We had more tea with the workers while the cab warmed up and then said our thanks and goodbyes. Some of the workers spoke a bit of German, like us. We learnt a bit of Turkish from them and taught them some English during our stay.

'Each time we went over the pass after that incident, we always stopped to say hello and on the return trips from Iran we stopped again with presents of Iranian tea which they really liked. It was always good to keep in with the grader men as we never knew when we would need their help.

'Once we were safely down the mountain and on the road towards Elazig, Bun reminded me that it was Christmas Day. We pulled over and opened the food locker for a celebratory meal. Unfortunately most of the labels had come off the tins so when we opened them we found the food frozen solid and ruined. All we had for our Christmas dinner was a cup of tea and a stale biscuit each.'

Gordon Pearce still has very clear memories of all the mountain ranges in Turkey which isn't surprising considering he completed over forty trips to Tehran. Without doubt the worst of them in his opinion was Tahir in the east of the country. It didn't matter whether it was spring, summer or winter, the hazards were always evident on Tahir.

In summer the roads were covered with loose gravel and the trucks would struggle to get good grip. The wind whipped across the mountain, making the

Tahir in summertime. Clouds of dust swirled everywhere especially around the trucks which were making the long, slow climb. (GP)

A totally different scenario. The same stretch of Tahir but in the depths of winter. (GP)

Gordon Pearce regularly drove BGH 172H. It is resting in Eastern Turkey en route to Iran in January 1971. (GP)

passage incredibly dusty and visibility very poor. Nothing could ever be kept clean and that included Pearce himself. 'It may have been three, four or even five days until I could have a proper wash. I looked and sometimes smelt like a tramp,' he said.

In winter, it was the hard-packed snow and ice and extremely low temperatures which were the enemy. Here is Pearce's account of a typical winter's trip over Tahir:

'January 1971 and I am en route to Tehran. It has been ten days since I left the UK. I'm driving a Scania drawbar outfit, BGH 172H, with a thirty-foot trailer in tow. The rig handles and pulls really well and everything seems smooth which makes the job so much easier.

'The twenty-foot container fixed to the truck is crammed full with twelve tons of razor blades. The trailer has around sixteen tons of groupage inside consisting of metal pipes, personal effects and helicopter spares. The all-up gross weight of the complete rig is around forty tons.

'Approaching the worst part of the route through Turkey and knowing the local terrain well, I have always stopped for the night just before the village of Horosan, before tackling Tahir the next day. Tonight is no different and I park on a garage forecourt which I know is safe. I shake hands with the owner and give him two packs of cigarettes as a way of thanks.

'There is a good foot of snow all around and my magnetic temperature gauge which I put on the cab roof tells me it's minus 15 outside. Inside the cab, it's much milder at minus 10!

'As I bed down for the night I can see the moon in a crystal-clear sky which makes it look just like a dull day instead of night. The snow glistens like millions of diamonds in the moonlight. It is a beautiful sight but also quite eerie as there are never any colours in moonlight.

'To help keep warm, I leave the engine running throughout the night. I set it at 1,200 rpm using the hand throttle on the dashboard. Letting it idle at a higher speed keeps the turbo blades moving and stops the mechanism from seizing up. Before leaving the UK, I modified the exhaust outlet pipe so that it now blows directly onto the fuel tank which is a simple but very effective way of stopping the diesel waxing up. The hot exhaust fumes also stop the air valves from freezing too.

'The night passes and in the morning I wake to find another clear blue sky. First thing I always do is make a cup of coffee, then walk around the rig to check everything is okay.

Pearce fitted his own hydrometer and temperature gauge to his Scania. The temperature inside the cab is minus 10! (GP)

'Then it's time to move off and leave the sheltered garage for the most exciting part of the trip. I always engage a low gear and move off slowly. The ride is a little bumpy to begin because all the tyres have frozen during the night. For the first half-mile or so they are still somewhat flat-spotted and very cold, but with some careful driving they soon warm up.

'As I approach Horosan the ride becomes much smoother and I feel more comfortable. I turn right in the village heading for Agri and after a few miles I drive through a cutting in the cliff, then begin to climb. This is the start of Tahir. A gentle slope for the first few miles makes for easy going and although the road is covered in hard-packed snow it is not yet bad enough for snow chains. Instead I drive carefully along the edge of the road for a better grip away from the tramlines left by other trucks.

In the gentle slope to begin the climb up Tahir, Pearce tried to keep out of the tramlines as much as he could. (PK)

'Two Iranian trucks approach from the opposite direction. This is a good sign as I know that the road ahead is clear and open. Another mile or so on and the road levels off so I pull into a lay-by and jump out to check all around the rig making sure things are still okay, especially the air lines, valves and tyres.

Hard-packed snow and ice. Pearce feared this part of Tahir the most. Snow chains were essential on the incredibly slippery surface. (GP)

'As the weather is worsening, I decide to fit snow chains onto the drive axle of the truck. It is bitterly cold so I have to work quickly. I jack up one side of the truck and wrap one set of chains around the two tyres.

'My fingers are tingling as I hold the ends of the chains in place and use one-inch strips of old inner tube with wire hooks on to join the ends together. I got the idea to use inner tube strips from Turkish drivers who swore by them. They work much better than the wire springs provided with the chains. The springs all too often stretch and make the chains useless as they keep slipping off the wheels.

'I let down the jack and repeat the process on the other side of the truck. It only takes about twenty minutes but I am definitely feeling the effects of the biting winds. Back inside the cab I make some more coffee before starting again to tackle the hardest stretch to the summit.

'With the cab heater turned up and the blower full on, my cold fingers soon return to a normal feeling but it reminds me of an incident last winter when I nearly lost my fingers. It was in this same spot when I was running in convoy with a Swiss driver who was on his first trip. We had stopped to check our trucks and fit chains. I did mine in twenty minutes and then went to help the Swiss guy.

'Unfortunately he was having problems with his chains, but I thought it was because he wasn't used to them. He had some fancy looking Swiss ones with cross links and another hook which went between the twin wheels. We struggled for a while to get his chains on but could not get the last parts on the outside of the tyre to come together. The pair of us tugged and pulled with all our might but it was no good.

'To make things worse the temperature was around minus 10 with a biting wind and my fingers had gone beyond tingling and were starting to go numb. I used my tyre lever to wrench the chains together but even that proved useless. I couldn't work out why they wouldn't close together so I picked up the box to look for a clue. I noticed the chain size was smaller than the wheels. No wonder they wouldn't fit!

'I rummaged around in my toolbox and found some spare chain links which I modified and then connected to the chains. I couldn't feel a damn thing in my fingers by then and just fumbled my way to fix everything together. It had taken us about an hour and a half but eventually we got the chains on and were ready to continue the journey.

'Before I started off again I put my hands in front of the floor heater in the cab expecting the numbness to go and the tingling to start but nothing happened. I rubbed my hands together for ages but still no feelings returned. I must have got a touch of frostbite that day because I didn't regain any feelings to my fingers for a good six months and then it was only a mild sensation. I counted myself as being very lucky that my fingers never dropped off!

'Back to the present: after enjoying my hot, strong coffee, I move off, coaching the truck gently along the road, picking the gears carefully so as not to cause the wheels to start spinning on the hard-packed snow and ice.

'I am soon down to second high in the ten-speed gearbox and running at almost full revs as I approach a really steep section of road. On my side of the left-hand drive truck is a thick, snow-covered, 45-degree upward slope. On the opposite side there's a 45-degree downward slope with no barriers and a sheer drop of some five hundred feet straight into the ravine!

'I continue to nurse the rig up the steep road taking care not to over-rev the engine and make the wheels spin. If that happens, I may not be able to gain control again so patience is the name of the game. Slowly does it. There are no trees here, just endless white mountains as far as the eye can see. It is so beautiful but I don't have time to admire the scenery, I must concentrate wholly on the driving.

'I continue on up for another half-kilometre as the road continues to get steeper and steeper. As used to it as I am, it's still a scary experience and one which I don't look forward to. Now I have to change down again into first high and the rig jolts and slows to just a crawling pace.

'I feel the wheels start to spin a little and know I am losing traction. I have to engage the diff locks on the move which is more difficult than it seems. Normally this should be done with the vehicle stationary but if I do stop, I definitely won't get going again. I quickly dip the clutch and flick the switch. Cross fingers and I hear the air-activated valve which tells me the diffs are engaged and have locked the two sets of drive wheels together.

'Another switch operates a valve which lifts the third, trailing axle of the truck, transferring all the weight onto the drive axle which also helps with traction. I am now concentrating hard and looking for areas of road with the best grip. Any side will do. Steady as she goes.

'I recall a previous trip on this very stretch of road. It was late at night and even colder at minus 30. I was unable to engage the diff locks and the drive axle lost all grip and started to spin. My immediate reaction was to push the clutch hard down, hit the foot brake, pull the handbrake on and then yank hard on the dead man emergency trailer brake.

'Everything stopped for a second and then I felt the whole rig start to slide … backwards! It was too dark to see the trailer in my mirrors and I suddenly sensed I was gaining speed. I began to panic, then BUMP! After fifty metres of fright, I hit something hard but at least I'd come to stop and hadn't gone over the edge.

'It turned out that the trailer had jack-knifed and ended up with its front axle turned almost right around. The truck had hit squarely against the side of the trailer which caused the bump. The rig had ended up in a T-shape. Luckily there was no damage to speak of, but I was unable to move the rig due to the angle which everything was at. I needed help.

'To add insult to injury, a couple of Turkish truck drivers were not happy with my predicament as I was well and truly blocking the road. They began shouting "Chok zug, chok zug!" which means very cold. Well, I knew it was bloody cold, but they meant hurry up before the diesel fuel waxes up and stops flowing to the engine. They were understandably worried that they would also get stuck.

'Within the hour a bulldozer from the nearby grader station came to my rescue and for the handsome price of 200 Turkish lira, about 6 pounds, I got a tow straight to the summit. I for one was glad to be alive and to live another day. I didn't want that experience to happen again.

'Back to reality, I am lucky on this trip as there is more grip on the road and I manage to keep the rig moving. The revs start to drop, so it's hard down with the throttle and I give it full power. The engine is working flat out and the turbo whistle confirms it. Then I pray that I keep moving as I watch the rev counter drop to 1,500 rpm.

'I have full torque but I dare not try and change gear just yet. I know this section of road well and very slowly the revs pick up again as the rig finds better grip. As usual, the butterflies leave my stomach and I feel more relaxed.

'Now there are just a few more hairpin corners left until I get to the summit. As the first sharp bend approaches, I disengage the diff locks and lower the trailing axle. This spreads the weight again and gives the front tyres a better chance of grip on the sharp bends; otherwise the truck could end up going straight on and right over the edge. It's a careful balance between wheel-spin and turning, watching and listening. The odd prayer here and there does not hurt, either.

'Finally, and with a huge sigh of relief, I reach the summit of Tahir. Disappointingly there are no real good views, just an open wilderness and bigger snowdrifts up to ten feet in places.

The summit of Tahir. Pearce was disappointed that he only saw more snowdrifts. (GP)

'I pull the rig up outside the grader station and go inside to see my friends. I take a glass of chai tea with them and chat for a while making good use of their roaring fire. I shake hands and promise to call in on the way back with some presents.

'On the move again, I ease the rig down the long descent which needs more care and attention than going up. The sun has melted the snow in places but the biting cold wind has frozen the water which was running across the road.

'Better to be safe than sorry, I decide it's best to fit a single snow chain on each of the front tyres of the truck and another chain on the rearmost wheels of the trailer, just to give better grip when trying to steer. It's so important to be aware of every situation and change in the road camber. The rig could easily slide sideways which would not be good.

'I notice some black smoke drifting in the air further down the mountain and I know exactly what it means. There is a Tonka on its way up and it will no doubt be grossly overloaded and chugging up the road at a very slow speed. At this altitude, there is less air for truck engines to breathe, hence the black smoke. Newer trucks like mine come fitted with turbos which make such a difference to engine power, especially in conditions like these.

'As the road is so narrow and mostly single track, I must quickly look for a wider stretch to pull into while the Tonka passes by because I know that he will not yield even though I am in a much bigger vehicle. Having done so many trips over Tahir, I am very familiar with the landscape and know the exact stopping place. When I stop and the brakes are applied, the rig may slide somewhat so I do not want to be stationary for very long at all.

'I frantically urge the Tonka on but the stony-faced driver just meanders past and gives a wave of thanks. I really don't like waiting around on the mountain for too long. It's a very formidable place and holds many bad experiences.

'On another bend further down, I come across a Bulgarian rig which has gone over the edge of the road and is lying on its side. It doesn't look good! The cab looks damaged, the trailer is split open and the load scattered. The most important unwritten rule on this job is never to drive past another truck which is in trouble without stopping to investigate. It may be you who needs help one day.

'It transpires that the Bulgarian driver and his mate had got stuck as they were coming up the incline. After trying to move again, their rig slid backwards ending up in a jack-knifed position, blocking the road completely. It was impossible for them to move.

Tonkas approaching. They wouldn't yield for any other traffic on the narrow single tracks. Note thick black smoke from exhaust. (GP)

'When the bulldozer came to rescue them, the men were asked to collect their suitcases and documents. They then stood and watched in dismay as the bulldozer pushed the whole rig over the edge of the road, effectively unblocking it!

'You may wonder why they didn't push my rig over the edge when I jack-knifed and got stuck on the previous trip I described. Tea is the answer to that, but it must be Iranian tea. The Turks love the taste of it far better than their own, so every time I come back from Iran I always deliver presents of tea at the grader stations. A kilo or two of it for the workers works wonders. Of course I have to stop with them for a drink as it's rude to refuse their hospitality.

'Now safely at the foot of Tahir, I make one last stop to remove the snow chains and pack them away in readiness for the next time I may need them. The weather is much better now I am clear of the mountain and I reckon I will make the Turkish exit border of Gürbulak today, so long as there are no more hold-ups.'

Gordon Pearce has always been an avid photographer who kept his camera and plenty of film in the cab wherever he went, on the lookout for unusual subjects. Over the years he has created an invaluable collection of photographs of the scenery, places and people he met and, of course, the trucks he drove.

He did, however, drive some of his workmates mad as he kept stopping to take snaps. Ray Scutts remembers many of these occasions:

'Gordon was a pain in the arse at times because he always wanted to stop and take bloody pictures of beetles and flowers. We'd have been driving for hours and were almost at our destination but Gordon insisted on stopping again to take another photograph. I'm sure he had shares in Kodak.'

Pearce's photography was so good that when the press chose to feature Asian Transport/Astran in various publications and magazines, they would often use his shots. Some of his striking images appeared on the front cover of the first edition of Franklyn Wood's book *Cola Cowboys*.

Pearce was asked on more than one occasion to take a reporter with him on his trips. The Automobile Association featured Pearce in the autumn 1968 edition of its members' magazine, *Drive*. The article was headed 'Hitching a lift with the Rambling Rose', an old nickname of his.

In 1971 he was featured in the *Daily Telegraph*

The Bulgarian Volvo F88 was bulldozed off the road, effectively unblocking it! Note Pearce's rig in background. (GP)

Magazine on another trip to Tehran. Pearce was accompanied in the rig by co-driver John Frost while the reporter and his cameraman followed behind in their car.

Through his relationship with the media, Pearce became quite a star in his own right which gave his bosses some very good publicity as all the articles portrayed him to be the perfect driver who kept his cool and dealt with situations professionally.

So why did Pearce leave Astran?

'I was never home more than a couple of days before I had to go off on another trip. In theory I should have had about ten days off in the UK but first I had to spend a couple of days of that time in the docks clearing customs. Then there were a couple of days doing deliveries all over the place, after which I had to take the truck up to Attleborough to get it serviced.

'I'd actually get home for just a couple of days before I had to go and pick up the truck and load again ready for the next trip. Eventually it began to wear me down, especially after I had done about forty trips in the seven years I was there.'

Bob Paul was very sad when his driving companion and friend decided to leave:

'He was someone to whom I had grown very close and I really didn't want him to leave my family. He was a very solid person who everyone respected and could rely on. He was greatly missed for a long time afterwards.'

After he left Astran, Pearce kept in touch with Paul although they did drift apart for some years, but much more recently their relationship has been rekindled. They clearly had a very special bond in the early days when they were literally living in each other's pockets for weeks or months at a time.

Both men have a calm, gentle nature, but also an adventurous streak. Above all, they share a great sense of humour, one of the main qualities which they needed to overcome some very difficult circumstances. Paul gave an instance:

'One of the funniest thing I shall never forget was the short back and sides haircut which Gordon slyly arranged for me.

'In one of the Turkish villages there was a barber's shop we used regularly and I decided I wanted a haircut that day so

Gordon and I both went in. It was a lovely warm place and the barber was very friendly and started chatting to us. I was absolutely knackered and after I'd sat down in his comfy chair, I promptly dozed off.

'When the barber woke me and showed me my haircut in the mirror, I was horrified. I was virtually bald!'

Paul's No. One crewcut became a standing joke for many years after that, as he always preferred to keep his hair longer. At the hairdresser's Paul had requested a light trim but after watching him fall asleep, Pearce thought that for a joke he shouldn't remind the barber. 'I could have killed Gordon on more than one occasion during our time together, but on that one I nearly murdered him there and then,' Paul confessed with a chuckle.

Forty-four years ago Pearce and Paul didn't know each other as they set out together on the first trip to Iran. During the journey their relationship sparked and turned into a very close friendship.

Today, the pair still share that special bond and it is lovely to see them reminiscing about an era of road transport that they were involved in, but which will never be repeated.

Thanks to Pearce's workmates – but not to Pearce himself – a final story about him can be revealed. Graham Wainwright and Eddie Parlane were intrigued by Pearce's habit of carrying piles of cocktail sticks and

Gordon Pearce was featured in many publications and periodicals. A selection of them here are Cola Cowboys, AA Drive *magazine and the* Sunday Telegraph Colour Supplement. *(CS)*

Bob Paul (left) and Gordon Pearce pictured in the winter of 1966 on their inaugural trip to Tehran. Between them is Uri, a friendly Iranian driver. (GP/CS)

British flags on his dashboard. Pearce didn't say what they were for, so Eddie asked his brother, Bun, if he knew. Eddie Parlane takes up the story:

'Bun replied, "I do know something. Keep your eye on Gordon and watch what he does with those little cocktail sticks and union jack flags. Over a period of time I wondered to myself where he kept wandering off to, as he had this habit of just disappearing."

'So next time Gordon and I were out together I waited until he came back to the truck after one of his wanderings. I'd seen where he'd come from and went off in that direction. I was very curious as to what he'd been up to. Then I came across one of his flags. He'd had a crap and stuck a flag in it. I couldn't believe my eyes. It was so funny.'

Gordon leaves a little bit of England behind, as imagined by Trevor Stringer. (TSC)

It transpired that Pearce liked to leave a little bit of England wherever he went! When Eddie returned home he told his brother who then confessed that he'd also followed Pearce into the woods and come across one or two of his little flags.

One of Pearce's favourite images. Refuelling at Macu, Iran, en route to Tehran. Mount Ararat and its little brother are clearly visible. (GP)

CHAPTER 13

Dick's Diary

The late Dick Snow's hand-written diary of his first trip for Asian Transport is a vivid account of the adverse conditions that the pioneering drivers met and overcame. Snow's widow, Cathy, remembers her husband going back to his diary in recent years and poring over them while he typed them up. He evidently added to his recollections and omitted some material until he produced the record that he wanted.

The following is based on his typewritten account. The content and the photographs of the typed diary pages are reproduced by courtesy of Cathy Snow.

First trip for Asian Transport
London - Tehran
Depart 13.12.1968 - Return 05.02.1969 (54 days)
Accompanying driver Raymond Scutts
Vehicle: Scania 110 drawbar outfit - WLO 95G

13.12.68

14.30 Left Brimsdown.

15.00 Arrive Brimsdown with lights and brake faults. Estimated time of departure 17.00.

18.30 Outside Brentwood, dolly nearside brake drums running red hot. Call fire brigade! (Approx 5 minutes, Brentwood and Chelmsford). Cool drums. Release all air. Proceed to Felixstowe with no trailer brakes.

23.15 Arrive Felixstowe. Meet Bob Paul and Mike Woodman. Had Cranes fix airlines, then board ship. Uneventful crossing.

14.12.68

10.00 Arrive Europort Holland. Wait in harbour for berth.

12.00 Berth. Unload and proceed to change tyre.

15.30 All well. Vehicle ready to proceed.

17.35 Outskirts of Rotterdam. Engine stalls and starter motor jams. Two very helpful Dutch drivers stop and help us investigate cause. Turned flywheel to no effect. Eventually one Dutch driver went home and got a rope to tow us.

18.00 We are proceeding again. Dolly brake drums quite warm but seem to be OK.

18.55 Leading rear axle starting to get very hot and smoking. Ray and I deciding what course of action to take with trailer brakes. Conclusion - Release leading rear-axle linings.

19.35 Released leading rear axle and proceed with caution. Seems to be pulling much better now.

20.05 Rear axle running red hot so have to release all air and run with no trailer brakes once again. Notice that rear brakes are very slow in releasing.

21.30 (approx) Arrive Elten/Emmerick. Ray's watch stopped at 20.05 so we are now guessing the time. Had a walk round the frontier and found not only was the border closed but also restaurants, so it was back to the 'Café Royal' for meat balls, Irish spew and new potatoes, followed by fruit cocktail and Carnation milk. Unfortunately our water container was leaking and so we were unable to make coffee.

Tomorrow we have a busy day trying to arrange some sort of efficient braking system so that we are ready to roll at 22.00 Sunday if we can get our papers cleared.

Speedo 771.

15.12.68 Sunday

Must write to Cath so that she has a letter from me before she goes north for Christmas.

09.00 Got up and saw our first signs of snow (half-inch). Weather quite cold but we are expecting it to get much colder as things go on. Made some coffee for Ray and started engine so as to warm the cabin.

After coffee, checked oil, water (low), battery and then off *for some breakfast. Ham and eggs cost 10.50 Dm. Return to truck and start work on trailer. Repaired offside parking light – bad earth. Jacked up trailer to check brakes. Found that rear axle drums were rubbing. Released and adjusted them. Now working OK. Leading axle adjustments made and all trailer drums now running free. We suspect that all trailer brake linings are breaking up due to rubbing a lot. If the trailer brakes now overheat we are going to proceed with no trailer brakes until we are clear of Germany.*

Sunday afternoon spent chatting in cab waiting for Emmerick border to open.

22.00 *Border open. Customs open 23.59. Cleared ourselves at border.*
  1. *Dutch customs*
  2. *Diesel card & tax*
  3. *Steuerkarte*
  4. *German customs*
  5. *Return to gate for clearance.*

16.12.68

00.45 *Finally cleared but blocked in car park by Dutch truck*

01.15 *Proceed into Germany Monday morning. Was a clear night. Starting to snow now.*

05.00 *Still having trouble with trailer brakes so have now released them right off and are running with no brakes. Vehicle seems to be pulling OK in this condition.*

Stopped at Seiburg approximately 100 miles from Emmerick. Snowing quite hard at times but tending to turn into sleet. Fill up at first Shell filling station 158 miles from Emmerick.

10.30 *Stop for breakfast at Sandort.*

11.00 *Proceeding. Weather clear. Roads damp. Uneventful journey apart from stalling engine.*

20.30 *Arrive Salzburg. Had meal and went to bed after adding more antifreeze. Expecting very cold weather from now on.*

17.12.68

06.00 *Customs await us like hungry lions. Procedure quite simple in fact. Almost the same as Emmerick border except*

Dick Snow about to depart on his very first trip for Asian Transport, 13 December 1968. (RS)

we have to have an Austrian road permit. Total cost crossing Sch.1684 and 18 Dm. Bloody expensive!

07.30 *Cleared. Customs very helpful so long as you know just a little German.*

We are parking up in Austria to telephone Mike Woodman and see if we can get something done about our trailer brakes and buy tyre levers and puncture outfit.

Bought most of the equipment we need and are now waiting for someone to come and look at our trailer brakes.

Spent most of the day having trailer looked at. Checked air lines, drums etc. Cleaned valve on Scania and adjust brakes. Seems to be working alright now.

Few miles later on road from Salzburg to Hallein brakes start to run hot but unable to stop because we are on a hill of sheet ice! Release drums at top of hill. Road conditions are getting bad and some trucks have snow chains on.

We are now having constant overheating of brake drums and release two rear axles leaving us with no brakes again.

22.30 Thought we passed John and Gordon. Pulled up but they didn't stop if it was them. Press on towards Liezen where we are going to park for the night. Both of us are quite tired now after watching for obstructions on this narrow road (159–E14–113–112).

18.12.68

01.00 Met John and Gordon who say conditions are bad further on and advise not to go past Graz until trailer working properly.

03.00 After exchanging chit-chat and Xmas greetings they depart. Gordon gave us forty cigs for Xmas.

Parked on mountainside in Austria and started at first light.

14.30 Arrived at Graz

Sent off cables to Asian and then proceeded into Jugoslavia where they charged us the earth for road tax! (£70.)

The weather today has been very bad. Constant snow and ice. Eventually we had to put our snow chains on as we got stuck in heavy snow at a garage.

21.30 Proceeded on our way from border only to find that the roads are sheet ice.

19.12.68

01.30 Stopped at Ormoz. Foot aching.

Checked tyres, oil and water. We had to top the latter up. On trying to start, we found the starter motor had either frozen or jammed and are about to investigate and then be on our way.

Weather bitterly cold with roads highly dangerous so will have to take it easy. The foot seems OK this morning after its ducking in icy water last night at the border.

Can get under 3.80 metres in Zagreb. Second lights turn left — Belgrade.

The weather has turned very mild and the roads are now clear of snow and ice. Ray and I are now driving as far as possible while the good weather lasts.

This evening it started to absolutely pour with rain almost like a summer thunderstorm and we both wonder if this is a good sign of better things to come in Bulgaria?

20.30 Still pouring with rain but now going much slower as we are driving on cobbles. We hope to pass Belgrade about 23.30.

23.00 Arrive Belgrade. Had a little bother finding our way round but made it in the end.

20.12.68 - one week out

10.30 Drove all night in four-hour stints and arrived at Bulgarian border. Stayed at border for a meal. Total of crossing was 73 leva which consisted of road tax 50, insurance 23.

The drive through Bulgaria was quite pleasant except for the first and last few miles or so and it took us a total of eleven hours driving. Roads well signposted and good surface except in villages where it is cobbles so you have to slow right down.

We arrived at the Bulgarian/Turkish frontier and have to wait until Sunday morning before making a move.

21.12.68 Saturday

The weather is very mild and we've had constant rain since Bulgaria. Last night we even had an electrical thunderstorm which seemed strange for this time of year. Speedo reading 2238 which means we have nearly gone halfway?

Wild dogs and cats at border.

Stuck at the border all day doing nothing much at all. Cleaned out the cab and washed outside. Met two Bulgarian drivers and had some of their Schnapps which was very strong. Ray declined the offer.

22.12.68

The agent says we will cross the border at 09.00 but the Turks are slow and we expect to cross about 15.00.

14.30 We are clear and ready to roll. Ray driving. Hungarian and Bulgarian drivers say not to stop at night in layby except in towns as there are bandits about on the northern and central routes.

The road to Istanbul is quite wide and has a decent surface for most of the way and one can get a move on. Our only trouble was that we caught up with about a hundred Leyland coaches travelling in convoy to Istanbul and Ankara. It was pointless trying to pass because the Turks won't let you in and in any case they were travelling too close together.

18.30 We arrived at Istanbul and stopped the night at the BP Mocamp. No sense in going on as coaches and military trucks have priority on ferry so we are hoping that we won't meet up with them again.

Speedo reading 2378

Most of the Iranian drivers we meet say the weather is bad in Eastern Turkey so we are getting prepared for worse things to come. It is now obvious that we are not going to make it for Xmas.

23.12.68

00.30 Went to bed

07.45 Got up. Had a wash and prepared breakfast (no bread since Salzburg).

Can't press on as we have to wait until Mocamp opens to see if there are any telegrams or letters for us. Rays says for first six days write to Mocamp and after that to the hotel in Tehran.

Ferry at Istanbul. Wait until you have ticket before proceeding on ferry or you will have to pay 250 lira instead of 170 lira.

10.45 Arrived at ferry. Twenty-minute crossing.

14.30 Got as far as Duzce and stopped for food.

Starter motor went on the truck. Tried for two hours to fix it but no joy.

Crowd of people gathered and one man asked if we spoke German. I said, Yes just a little and managed to convey to him what was wrong. As luck would have it there was a chap 100 yards down the road who specialised in electrical starter motors and they said they would try and mend it for us. They were very helpful indeed and worked from 18.30 until 08.00 next morning. It cost us the earth but was well worth it. We were half-way between Istanbul and Ankara and it would have cost more to have an agent come out and repair it. It seems to function as good as new so there is much to be said for Turkish labour efforts.

24.12.68

11.30 Thought we were well on the way to Ankara keeping a steady average speed but alas once again we had to stop because of a puncture on rear axle of trailer. No spare tube so about turn and back to Bolu to search for a bank and tyre garage.

Surely this must be our last fault before Xmas which is tomorrow. No doubt we will spend it stuck up some mountain with some fault or other.

Ray is just about fed up with constant faults with this new machine and is almost on the verge of flying home from Istanbul/Ankara. It's OK for me, this being my first trip, but for him it was the same on his last trip with the AEC.

20.10 Arrived Ankara and decided to press on.

Wednesday 25th December – Christmas Day

```
27.12.68.
07.00 – Checked oil, water and tyres and dropped radiator fan.  Everything
working.
07.25 – We are away.  Thick snow on roads but no wind to drift it.  Sky full
of snow.
08.15 – Stuck on hill again – it's not having double chains that is doing it.
08.55 – Snow plough tows us up hill once again at the cost of another 40 cigs.
We are now on top of mountain (1980 mtrs.) and are descending as slowly as
possible.  Snow plough following us down.  All seems to be going well.  Snow
getting very scarce and roads almost clear.
10.00 – Once again came upon steep hill and have same trouble with outers
gripping and inners slipping so stuck again.  We are going to try putting tow
chain under one inner to see if this helps us out of this predicament.  If not
we will have to wait for snow plough to arrive and tow us up once again.
Speedo 2981.
11.15 – Our attempt was unsuccessful so now we are waiting for the snow plough
to drag us up to the top.  We are going to buy a set of 1100 x 20 doubles at
the nearest town and hope that our troubles are over from then on.
12.00 – Snowplough arrived and the bastards wanted 20 cigs each and half a
tin of coffee !  What else could we do but give it to them?  It's a matter
of pay or stay put.
```

The typed version of Dick's diary.

03.15 Parked for meal 30 km south and had long chat. Next thing we knew it was 03.00 on the 25th so decided to sleep where we were and press on in morning.

Went to bed after wishing each other HAPPY CHRISTMAS.

Breakfast. Black coffee and press on to nearest water hole for nice cold wash.

Weather fair. Wind north-east Force 6. Roads dry. No signs of snow.

10.00 Up and away.

Speedo reading 2691.

We are making good progress today and have decided to carry on through the night to Malatya which is about 160 miles further on.

20.00 Pulled up for meal and were informed that the road was blocked by a bus so had to park up.

26.12.68

A nice sunny morning with a very strong wind. There seems to have been quite a bit of snow in the hills during the night but the roads are dry and clear.

07.15 We push on.

09.15 Wind Force 11/12 with snow blowing off mountains making visibility zero.

09.45 Blowing a blizzard and it is dangerous to get out of the cab for fear of being swept away.

We are now parked behind two Turkish lorries and are waiting for it to blow itself out.

Both in good spirits. The only thing we are worried about is that the trailer has no brakes and if the wind increases it might move the trailer across the road and we don't know if there is a steep drop or what on the other side.

Speedo 2954

14.30 We were stuck here until now when the wind abated and snow ploughs came along.

16.46 Had chains on and managed to get the engine started. It had frozen completely on the starboard side. This was with engine running at 1100 revs.

17.06 We are moving slowly. Three minutes later got rear wheels in pothole and couldn't move. Blocked road completely. Turkish soldiers and civilians came along to give us a hand but unable to roll.

21.00 Grader personnel eventually towed us up the hill and we parked at grader station having moved a total of approximately twelve miles in twenty-four hours.

It must have been really cold to freeze a Scania with the engine running! The two Ford D series in front of us kept their engines going all the time.

Speedo 2959

27.12.68

07.00 Checked oil, water and tyres and dropped radiator fan. Everything working.

07.25 We are away. Thick snow on roads but no wind to drift it. Sky full of snow.

08.15 Stuck on hill again – it's not having double-chains that is doing it.

08.55 Snow plough tows us up the hill once again at the cost of another 40 cigs.

We are now on top of mountain (1,980 metres) and are descending as slowly as possible. Snow plough following us down.

All seems to be going well. Snow getting very scarce and road almost clear.

10.00 Once again came upon steep hill and have some trouble with outers gripping and inners slipping so stuck again.

We are going to try to put tow chain under one inner to see if this helps us out of this predicament. If not, we'll have to wait for snow plough to arrive and tow us up once again.

Speedo 2981

11.15 Our attempt was unsuccessful so now we are waiting for the snow plough to drag us up to the top. We are going to buy a set of 1100 x 20 doubles at the nearest town and hope that our troubles are over from then on.

12.00 Snow plough arrived and the bastards wanted twenty cigs each and half a tin of coffee! What else could we do but give it to them? It's a matter of pay or stay put.

13.15 Arrived in Gurun. Two Hungarians say road is clear so we press on.

Twenty miles on, the brakes start smoking so we pull up and let them cool off a bit while we have our lunch.

14.30 Weather: clear skies and sunshine.

21.15 At last we have arrived at Malatya.

Speedo 3076

Luckily the shops are still open and Ray has gone off to buy a pair of 1100 x 20 double-chains. I hope he is lucky because I don't fancy having another three days of being stuck on mountains and having to wait for snow ploughs to tow us up.

We are going to park up here tonight as we are both shagged out after concentrating all day driving up and down hills of sheet ice and snow!

I seem to be a star attraction with blond hair and red beard. They can't figure this at all.

28.12.68

07.00 Woke and checked oil, water and tyres. One puncture. Blast!

09.30 Ready to go. Ray off to bank for more money for snow chains. It will just be our luck not to have any more snow. Still, we need them badly for the return trip.

14.30 No luck. On to Elazig and arrived with no hold-ups. Perhaps our luck has changed?

29.12.68

06.30 Drove all day and night and were as far as Adilcevaz where we had to stop and repair yet another puncture

08.00 On our way again in brilliant sunshine and clear roads. We did have a little mud but it didn't give us any trouble.

We stopped at Lake Van for a wash and I was tempted to jump right in but had second thoughts!

11.30 Passed through Ercis and things started to look a bit grim as we were now running on snow again.

12.00 Stuck fast on hill with nearside wheels in ditch. Awaiting bulldozer to drag us clear.

While waiting we had our first meal of the day – corndog, beans, carrots, mushrooms and are now waiting for a saucepan of snow to melt for our tea.

Both in good spirits although with constant breakdowns and snowbounds we shouldn't be. Have decided possible cause of trailer brakes locking to be either protection valve or two-way valve on Scania. We are losing a lot of air on the front gauge.

Still hope to make Tehran for New Year with luck and no snow.

13.00 Still waiting for bulldozer. Almost warm enough to sunbathe?

14.00 Still waiting. Am now stripped to waist. Weather brilliant sunshine but road still snowbound.

15.00 Graders arrive in truck and demand cigarettes. We say 'Cobblers' so don't know what will happen now! Have tried everything but the only way up is double-chains. We will definitely not return without them.

30.12.68

06.30 Freezing hard and still no sign of bulldozer.

We review the situation which has slightly worsened during the night. We find that our air lines are frozen although we drained them yesterday when the sun was on them.

09.30 Graders go up to station and once again say 'Yes,' they will bring the bulldozer. We try once again to help ourselves but seem to bog ourselves in further so will just have to wait and rely on them.

Sun up and very warm but no sign of a thaw.

11.30 After digging with a screwdriver and hammer for two hours, I managed to clear a patch for the nearside wheels to get a grip for approximately six feet and then we were away – only to meet the bulldozer coming! Hard luck. We waited for two days and then are away.

14.15 Forty kilometres from Agri we meet four Hungarians who say it has taken them two days to get 40 km and that an English motor is stuck 10 km down the road.

15.00 Passed Turkish Transport and Trading ERF stuck on hill. Looks like he has been there for a few days! No sign of life. Approximately 2 ½ miles from Hamur.

15.55 Agri at last. The last sixty miles have been a mixture of sun, mud, ice, snow, fog, ice and snow. Thought we were never going to make it. Let's hope it's OK up to the border.

Saw a small brown bear hanging around a BP station.

21.00 Arrived at Gürbulak and crossed into Iran. Had to wait until morning for customs and agent.

31.12.68

Hope to clear customs by mid-day and arrive Tehran some time tomorrow.

Met all the Hungarian and Bulgarian drivers who went different routes through Turkey and weather was much the same because we all arrived at the border within a few hours of each other so we haven't done too badly!

The usual waiting period and arguing about who is next on the loading bank. We are supposed to wait until tomorrow.

Opened my bottle of whisky this evening and had a little party amongst the drivers. Ray declined and went to bed not feeling too well.

Wednesday 1st January 1969

01.00	Bed.
09.20	They start to check our trailer.
13.20	Finished trailer check.
14.00	They start on truck.
16.30	Everything is completed and we are ready to move once more. By now Ray is not only feeling ill, he looks ill! I will drive round the clock.
22.00	Stop for Oxo at Tabriz and then press on.
23.59	Stop for sleep.

2.1.69

01.30	On again.
03.00	Have to stop for sleep.
07.30	Daylight and refreshed. Push on once more.
	The trailer is swinging about quite a bit and I keep checking tyres but all seems well. Perhaps it is the dolly because it is swinging at front, not rear.
08.39	Stop for coffee. We are now nearly half-way to Tehran. Ray still sleeping.
16.30	Arrived Tehran after killing three dogs on outskirts!
18.00	Arrived hotel for well-earned rest.

Follow signs for Tehran South – Diesel Whites. Roundabout. Turn left. Up on right. Leyland's roundabout turn right. Up on left.

Customs: Down Ghazvin to roundabout. Turn right. Straight past station. 1st lights turn right. Down on right past car park. Levant Express opposite customs entrance.

Asian Transit: Avenue Ferdowsi.

Continental Hotel: Up Ferdowsi. Over island. Up past Marmar hotel. Cross two lights. Turn left and across one junction. Next to Cleopatra club.

Main Post Office: Down Ferdowsi. 1st right, on right.

British Embassy: Avenue Ferdowsi.

16.1.69

Hotel bill 2.1.69–16.1.69 inclusive £146 for two persons (25555 Rials)

Bob says we shouldn't have paid 15% charge but no one told us so of course we paid. Next time we'll know different.

17.1.69

Speedo 4130

10.00	Left Tehran. Ray driving.
11.30	Brakes on fire thirty-eight miles out. Blow out.
14.00	En route border. Fine sunny day.
21.00	Met the two Bobs [Bob Paul & Bobby Vallas]. They left UK 1.1.69. Had some of their food and booze and best of all 90 Dm.

18.1.69

00.30	Stayed chatting until now. Info to go through Patros easier way.
03.30	Arrived Tabriz. Road a bit icy but no trouble at all.
	Speedo 4500 not bad going.
	Had a brew-up and first feed of the day thanks to Bob.
04.00	Bed.
07.30	I woke.
08.00	Washed and away. Roads dry and weather sunny. No snags.
13.05	Arrived Bazargan. Hope to clear customs and do an adjustment to clutch and also repair chains. Drivers say weather very bad in Turkey.
16.15	Cleared border. Roads very icy. Chains repaired.
21.00	Stuck on hill. Put towing chain in pothole and managed to drag ourselves out.
Midnight	Arrived Agri. Roads getting worse.

19.1.69

Changed money and proceeded down wrong road!

10.15	We are now on wrong road and are unable to turn round. Much trouble with locals! Roads extremely difficult to pass but are keeping on the move at approximately 25 mph. Almost got stuck at Adilcevaz.

Very steep hills covered in snow and ice all the way up. Good driving on Ray's part saved the day, thank Christ.

19.30   *Arrived Tarvan. Roads not clear of snow so parked up. Snowing hard.*

20.1.69

06.30   *Graders have cleared road so push on.*

08.30   *On wrong road (Bitlis). Turned round and then got stuck on hill. Bulldozer assisting snapped chain and then pulled bumper away from chassis. Managed to get wire hawser round axle and he then towed us up.*

14.00   *Having fuel trouble. Checked everything but we are still getting air in somewhere.*

16.00   *Still same problem but are unable to solve.*

17.30   *Dark and unable to travel. No lights due to bumper.*

21.1.69

Check lines from tanks but can find no leaks of any sort. Bleed system but air still getting in.

08.06   *Stroke of luck – saw very tiny bubble coming from flexible hose leading from main filter to pump. This must be our trouble. Police came along and said they would run Ray back to Mus to see if he could get a new one or one made up.*

09.45   *Ray still not returned. Hope he is having some luck.*

12.30   *Ray returns having walked five miles in snowstorm, looking rather bedraggled. No luck but managed to get some food and jubilee clips so we try those. Still no joy.*

14.00   *Have disconnected main fuel lines and connected air line. Still no joy. Drained all filters and bled. Still nothing. It seems there is no diesel getting to Numbers 1 and 2 injectors. Hope it's not the pump.*

17.15   *Darkness coming on fast so have to abandon all attempts for today. Sat up reading manual which gives lots of suggestions for workshop experts with tools but with limited facilities we can't do a lot.*

Apart from that we are up to our knees in muck and snow which makes work limited to a couple of hours before we have to warm up. Talk over various ideas of getting help but are stuck in a bad spot, Malatya being 190 miles away and nothing to the east except Tehran. We are hoping that John and Gordon will come along to at least buck up our morale which is getting a bit low after following the instructions in the manual to the letter and still getting no joy.

22.30   *Had visitors trying to force the window, but they disappeared on shouting 'Piss off, you Turkish bastards!'*

22.1.69

06.30   *Up and had breakfast and decided to check from pre-filter to injectors. If there is still no solution, Ray is going back to Tehran (832 miles).*

There is no PTT at Mus so we cannot contact anyone east or west. What a place to break down!

17.00   *Talked over various remedies but came to no firm conclusions. If Gordon arrives, Ray will go back to Tehran.*

17.30   *No cigarettes and little food and then a ray of hope comes galloping round the corner: Gordon and John. Food, cigs, warmth – morale 100%!*

Talked over the problem and decided to have a go again tomorrow. Four heads better than two.

23.30   *Played cards and went to bed.*

23.1.69

06.30   *Up and had coffee and started to work.*

08.30   *Changed pipes all round and found they worked. Managed to clear and get engine going at last. Our problems solved.*

Collected chains, food and cigs from Gordon. Good lads, saved our day.

12.00   *Had our Xmas dinner – chicken, spuds and carrots. Sat chatting and reading.*

13.00   *Still at it and have decided to have our bumper repaired at Bingöl and park for the night. Hope we are now over our problems.*

15.00   *Said our fond farewells and went our separate ways.*

A few miles down the road the main warning light flashing. Checked oil, water, air and flashers – all OK. Must have disturbed a wire while we had the cab up so are pressing on.

21.00   *Stuck on hill. Double-chains OK but locking mechanism broken. Park for night and get some rest.*

Travelled 44 miles.

24.1.69

06.15   *Up and started working on chains.*

07.30   *Managed to clear hill and proceeded. Snowing hard.*

08.30   *Stopped for breakfast and repair to broken link.*

It is much better now that we have double-chains as we can help ourselves out of most problems. Had to put on the chains a couple of times during the morning but managed to climb all hills.

14.00 *We are running on dry roads.*

14.30 *Met Hungarian driver and exchanged road conditions. He says roads clear to Malatya but 25 km out it starts to get bad again.*

Took a message for his comrades who are somewhere behind.

17.38 *Met rest of Hungarians and passed on message (in German). These chaps are very helpful in telling us where we can get repairs etc. It's very nice once again to meet international drivers – there seem to be no barriers at all. Very friendly lot indeed.*

18.30 *Arrive Malatya and parked for the night. Must get our chains repaired before pushing on over mountains.*

Cable Mike Woodman for German permit and Austrian permit. – also more cash.

If we have more breakdowns we will not have enough money for Yugoslav and Bulgarian road tax and insurance.

25.1.69

07.00 *Up and did lots of repairs to chains (doubles). Try and get bumper fixed so we have lights for night driving after Istanbul.*

10.30 *Chains all fixed. Put oil and water in engine. Couldn't obtain meths for brakes so hope they don't freeze on over the next two mountains.*

Ray still driving and says he is going to drive all night until he is tired. I should worry, I'm getting paid just the same.

Roads quite good. Had to use chains twice and then almost clear roads to Gürün where we are stopping the night. Turkish drivers say very bad hill ahead and best to pass over in daylight so we take their advice.

Travelled 89 miles in six hours. Chains holding up very well. Turk made a good job on them.

26.1.69

08.30 *Left Gürün with chains on. Heavy snow in the night. Road conditions hard-packed snow.*

17.30 *We manage to keep a steady 20 mph and arrive at Kayseri.*

Only one snag today: the diesel warmed up again, so had to attend.

Roads seem to be clear from Kayseri so we are pushing onto Kirsehir and possibly Ankara.

19.00 *Stop for coffee at 1,260 metres. Supposed to be the highest point on this road but once again we are on packed snow. Hope we don't have to put the chains on.*

21.45 *Went to bed.*

27.1.69

01.30 *Woke and are just entering Ankara. We are pushing onto Bolu where Ray must cash some more money for the ferry.*

07.30 *At Bolu and having trouble with bogie bolts all falling out. Will have to wait until 09.00 for banks to open so have breakfast. Our best day since leaving Iran!*

Roads look bad ahead but we hope we have passed the worst yesterday. Hope to arrive Istanbul early this evening, roads permitting.

08.45 *Left Bolu but bogie is still not working. I think perhaps it is frozen up. Forty miles further roads seem clear and we take chains off. Bogie now working.*

One hour later put chains on for about 500 metres and then take them off again. Roads now clear.

Fifty miles from Istanbul now running on black ice.

19.30 *Arrived Istanbul. Black ice all the way.*

Had our first bath for eleven days and received mail from home at long last. Should be well into Bulgaria tomorrow night, weather permitting.

28.1.69

07.00 *Up. Weather foul, snowing like hell and black ice on roads.*

10.30 *Proceeding with extreme caution.*

17.00 *Arrive Turkish/Bulgarian border.*

17.30 *Cleared and press on for Yugoslavia. Roads alright but lots of black ice mainly in towns.*

29.1.69

03.00 *Arrived Bugarian/Yugoslavian border.*

03.30 *Cleared.*

04.30 *Press on for Austria. Customs say roads are clear so we are hoping for a quick crossing. Having trouble with clutch* pedal. Keeps going solid. Quick clear seems to be to drain air tank when it is not frozen up.

10.30 Arrive in Belgrade having covered 564 miles in the last twenty-four hours. Autoput [motorway] dry and we are now going like a bat out of hell. With this average we should be at Austrian border early morning tomorrow.

20.40 Surprise, surprise. Arrived at Sentilj and proceeded to cross. Roads OK Yugoslavia.

22.30 Crossed border and carried on across Austria.

30.1.69

06.00 We are running into snow but don't think we will have to put chains on. Should arrive at Salzburg around 13.00.

12.30 Arrived Salzburg. No permits for Germany but letter to say we can draw up to £100. Just waited for rest of day doing nothing much.

18.30 Bed.

31.1.69

Still no permit so hang about border all day. Cleaned out cab and then went into Salzburg and had our bumper straightened out a bit so that we would not have any trouble with the Germans. Agent says permits may be here tomorrow. Let's hope so!

Going to bed early again tonight in case we have to drive right through Germany in one go. It can be done in about fifteen hours.

1.2.69

Still at Salzburg awaiting arrival of German permit. Check truck and trailer for faults and rectify if possible. Not much else doing except eating, drinking and then sleeping!

We spend most of the day chatting to Tekchm drivers [Turkish transport company].

2.2.69

Much the same as yesterday.

3.2.69

10.00 Permit arrived and we proceeded to cross border.

13.00 Completed. We are now going like hell for Rotterdam trying to catch tomorrow's boat.

4.2.69

03.30 Cross into Holland.

04.00 Clear.

07.30 Arrived boat.

12.00 Sailed.

18.20 Arrived Felixstowe

5.2.69

01.30 HOME AT LAST!

1.2.75
YILDIZELI, SIVAS, TURKEY

Eighteen tyres cheeked and singing on the pavement. Diesel smoke blowing in the wind. Five thousand miles to go; six weeks away from home and Mister that demands the BEST.
Snowy -40°c Sivas Turkey

CHAPTER 14

Slips, Trips and Falls

Considering the colossal amount of miles that the men drove each year, it comes as no surprise that a fair amount of slips, trips and falls occurred. And being in the wrong place at the wrong time could bring very serious consequences.

Geoff Frost nearly ended up in a difficult situation through no fault of his own in Syria. He had set off from the border at Bab el-Hawa and late one evening, just after he passed through the small northern town of Hama, an old Mercedes car pulled in front of him. Suddenly someone opened the back door and threw a body out!

The aim of the criminals was to make it look as though the body had been hit by a truck, but Frost's reaction was perfect and he managed to swerve to avoid it. He didn't stop to investigate and he couldn't tell whether the body was dead or alive.

Frost would have been blamed for the accident and would have ended up in jail for a long time. As Frost said, 'In the Arab world, you don't stop for anything because in a case like that it comes down to blood money and you just don't get involved.'

Gavin Smith also remembers Syria well:

'The trucks and coaches had lights which barely glowed and they had red ones pointing forward. I remember coming home in the convoy system. We were on a new bit of dual carriageway and were all heading north near the town of Homs. I was happily buzzing along empty, when I saw a faint glow getting closer and closer, so I put on my high beam to get a better look and was promptly 'microwaved' by a huge set of lights that were attached to an old truck coming towards me ... on my side of the new dual carriageway! Work that out.

'I took evasive action and almost had to change my pants. Luckily, I made it to Bab el-Hawa in one piece. Driving in the Middle East was like that but it was all in a night's work and something we got used to.'

Examples such as these could be found throughout any trip to the Middle East as it was the way things were, but as Frost concluded: 'I was involved in one or two altercations in my time. However, we all knew the risks and we just got on with the job and kept our noses clean, so we were generally okay.'

Being the largest country to negotiate and having the worst weather conditions to endure, Turkey bore the brunt of most slips, trips and falls. All the drivers have vivid memories of the forbidding country and its numerous hazards.

Without doubt the Astran driver who had the worst fall was John Williams. In the late part of 1974, he set off from Addington in his Scania drawbar outfit, JLL 686K. He then became stranded on the formidable Tahir mountain pass in Eastern Turkey as he unselfishly tried to help two other, non-Astran, drivers who had become stuck. Unfortunately he could not stop his own rig from sliding into a ditch as he went to their aid.

Dick Snow, en route to Iran, came upon Williams after the incident and stopped to help his colleague as best he could. Williams described the events during the BBC television documentary *Destination Doha* and it is clear from the expressions on his face and the tone of his voice that the incident affected him greatly. Here are some of his comments taken directly from the film:

'I had made it over the pass and dropped my trailer to go back and pull two of my friends who were stuck because this is a really powerful truck. Half-way down the hill, I came across three trucks that had jack-knifed across the road. While I was trying to creep around them, I slid off the road and she went over at about 45 degrees.

'It was during the Cyprus troubles and there was a lot of anti-British feeling in the Turkish army, so I couldn't get a crane or a bulldozer or anything. It took me a month and I jacked it out piece by piece, inch by inch. Some really good lads stopped and helped but I was a month with this just so no one would touch it. Not even a mirror.'

Martin Sortwell, a good friend of the late Dick Snow, filled in the details of the story:

'Snowy and I were on our way to Iran and had been advised of Willy's incident by telex as we came through Austria.

'It took us about a week to get to Tahir and when we saw Willy's truck we instantly knew it would be a long hard slog to recover it. It turned out to be a horrendous job trying to

Dick 'Snowy' Snow (left), John Williams (centre) and the Scotts driver make fun out of the serious situation. (MS)

get it out of the ditch and the whole process actually took about three weeks which is incredible when you think about it.

'Snowy and I stayed with Willy for the last ten days or so of his imprisonment in the ditch, but there were many other drivers who also stopped to help at some point. There were some Hungarocamion drivers who made the most wonderful coffee-cognacs and a chap from Scotts Transport of London who had dropped his trailer further along and had returned with his tractor unit to try and help.

'We all took turns digging, but there was granite under the

Snowy supervises the digging which was almost impossible at times. Note the number of days that Williams had been stranded, written in the dirt on the truck. (MS)

Slips, Trips and Falls

Williams used two jacks in an attempt to push the truck out of the ditch inch by inch. One of the jacks is just visible to the right of the picture. (MS)

snow and ice where the truck was lying which made it extremely difficult. It was just too hard to break up. During the night the temperatures plummeted to around minus 30 and of course the next morning everything was absolutely solid so we had virtually to start again. It was so demoralising.

'We only had very basic tools like chisels, hammers and a couple of long-handled shovels, so you can see why it took so long. We lit some wood and placed it in the ditch to make digging in the frozen conditions easier, but decided that wasn't such a good idea when the tyres nearly caught fire.

'Willy even used his wheel jacks to try and push the

Left to right: John Williams, Jeff Ruggins, Snowy, Dick Rivers and the Scotts driver (just visible) kept warm by the fire which was constantly burning. (MS)

Martin Sortwell (left) and Williams (standing in cab) celebrate the 'release'. Snowy's rig is parked behind. (MS)

truck away from the bank inch by inch so we could drop some rocks under the wheels to try and level it up.

'It was so bloody cold at night we used to have a fire permanently burning to keep everyone warm. We used the wooden pallets and boxes from the trucks that had jack-knifed and I remember frying eggs on a shovel which I held over the fire.

'We also had plenty of booze because everyone who went past used to leave us something. There were all nationalities so we had vodka from the Russians and schnapps from the Bulgarians. We ended up having a party and getting pissed every night which seemed to make the situation much more bearable.

'With the truck leaning at such an angle we had to think about sleeping arrangements. We all piled into Dick's cab for a night cap. There were two bunks in our Scania but only one in the Scotts Volvo. I remember Willy used to sleep bolt upright in a seat when he went to bed.'

Jeff Ruggins from Astran and Dick Rivers from British Leyland, sharing a truck, were returning from Tehran when they came across the scene. They were trialing a new Leyland Marathon which had double-drive axles, making it well suited for the rough terrain. Sortwell continued:

'We had the idea of chaining up another truck alongside Willy's and trying to pull his truck over just enough to get his wheels on a more level angle on top of the rocks. The Scotts truck was not man enough so they used the big Marathon instead. By this time, there was also a bulldozer which had been begged from a local grader station.

'The bulldozer's driver got behind Willy's truck and pushed, while Ruggins kept the chain taut and pulled forwards at an angle from the side. Willy then managed to steer his truck and up it came onto the rocks and then out of the ditch and onto the road. It was due to Ruggins's expert driving that it came out, otherwise it would have carried on along the ditch on its side.'

Sortwell said that the relief on Willy's face was unbelievable when he was finally released from his four weeks of hell.

Gordon Pearce had many slips, trips and falls during his seven-year career with Asian Transport and Astran but, ironically, the only serious one was when he tipped the whole rig over! Luckily no one was badly hurt.

The AEC rig which Pearce tipped over. Amazingly it was re-built and completed many more trips to Iran. (GP)

However, to make things worse, he was accompanied by his boss, Bob Paul, who still suffers some neck problems as a result. The incident dented Pearce's pride more than anything else:

'We were coming back from Iran and I was only doing about twenty miles an hour driving at night. We were on an approach road to the Tahir mountain pass and Bob was sleeping on the bunk. A slight bend came up with a camber falling to the left and then suddenly I noticed that the road was covered in mud.

'Next thing we skipped and skidded across the rough corrugated surface and straight off the road. There was nothing I could do to stop it and the AEC landed on its side.

'Bob woke up standing on his head in the corner of the cab. He'd hurt his neck but I was okay. We'd ended up in a field and the trailer roof had ripped right open. The load of carpet bales was all over the place but luckily the bales were all intact.'

As daylight appeared, Pearce opted to stay and guard the truck while Paul went to get help. After he had contacted the agent back in Tehran, he managed to muster up around ten local Tonkas to come and pick up all the bales and take them on to Basle in Switzerland which was as far as they were allowed to go. From there, the carpets were loaded onto rail wagons and sent to the UK.

Paul flew home to oversee the final part of the journey and make sure the carpets arrived at the customer complete. Meanwhile Pearce eventually managed to get the rig back to Tehran for repairs:

'Bob had arranged for a crane and a bulldozer to pull the rig out to the road and then I actually managed to drive it, even though the cab and chassis were twisted and the trailer roof was hanging off. Actually, there was no other feasible way of getting it back to the Leyland garage in Tehran so it was lucky that it was driveable.'

The truck batteries were mounted inside the cab underneath the passenger seat and while the truck lay on its side in the field, they began to leak and cover the men's clothes in acid. The weather was warm and after a few days Pearce noticed that his clothes were starting to fall to pieces.

As it was going to take quite a while for the AEC to be repaired, Pearce decided to travel back to the UK on a coach:

Pearce had warned his pupil to slow down, but his advice went unheeded and the new boy tipped the trailer over. (GP)

'I started off in an Iranian one which went all the way to Munich. Then I got a German one to the ferry port at Rotterdam and hopped on a ferry to England where Bob met me and drove me home. The journey took seven days and it was very nice to relax, enjoy the scenery and let someone else do all the driving.'

Meanwhile, back at the British Leyland garage in Tehran, the mechanics stripped everything down from the truck, got the chassis straightened and then rebuilt it all back to new. Chief mechanic Harry Sully and his team did a wonderful job.

To celebrate the rebuild, the Iranian mechanics fitted many different-coloured lights all over the AEC. The local trucks were all decorated with these bright illuminations as a religious symbol and good-luck charm. 'Although they all looked very pretty at night and I'm sure they served their purpose, they weren't really what we wanted, so I pulled them all off when we got back to England,' Pearce said.

As number-one driver, one of Pearce's duties was to take new drivers on their inaugural trips to show them the ropes and, more importantly, to determine whether they were up to the job. Pearce clearly remembers one new boy who was certainly not cut out for it:

'You needed patience and a calm nature when driving across unknown territory. We'd only just started out on a trip to Tehran and I wasn't very impressed with his driving techniques. I had to stay awake and keep an eye on him as he drove us through Germany.

'As we continued through Austria we were following the river and the road was quite narrow and very twisting in places. He hadn't improved much since leaving home and I just couldn't tell him how to drive. He didn't want to listen.

'Then at one particular tight S-bend he couldn't slow the AEC down enough and swerved across the road right on a level crossing. The whole rig was swaying badly from left to right. Luckily, the truck stayed upright, but the trailer tipped and landed on its side.

'It was fully loaded with a large packing crate full of furniture and boxes of whisky which were packed tightly all around it. To be honest I thought we'd lost the lot but amazingly only three bottles were broken. We had packed them in so tightly that nothing moved.'

Pearce made contact with base and after a couple of days Michael Woodman and Bob Paul arrived on the scene in the company Minivan. Woodman gave the other driver enough money to make his way home, then sacked him on the spot. He drove himself home in the

Luckily there was very little damage done to the rig or its cargo. (GP)

Mini, while Paul stayed with Pearce to recover the truck and continue the journey to Iran.

First they had to hire a crane to help unload the trailer. It was then up-righted and, after repairs were made to its torn roof, it was reloaded and re-sealed. The whole incident cost Pearce around four days. As to what happened to the sacked driver: 'I don't remember and I don't care,' Pearce said bluntly.

Pearce also had some very bad luck in Iran when he was driving a brand-new Scania Vabis on its first trip:

'I was driving while my mate John Frost was resting. We were travelling quite nicely on a freshly graded stretch of road near the town of Zanjan. The truck was fully loaded and wasn't too heavy, but the low-loader trailer had an oil-filled electrical transformer on it which weighed about 25 tons.

Michael Woodman (hand on head by Minivan) and Bob Paul arrived to supervise the recovery. (GP)

Pearce's predicament caused much amusement amongst the others. Left to right: Bun Parlane, Eddie Parlane, Graham Wainwright and John Frost. (GW)

'It was night time and I could see the headlights of another vehicle coming towards me. Before I knew it, the oncoming vehicle was virtually on top of me and directly head on so I had to swerve sharply to avoid a collision. The edge of the road was soft and as the trailer wheels ran into it, the weight of the transformer combined with the angle of the verge dragged the whole thing downward.

'The rear end of the truck was also pulled down and I ended up in a very precarious position with the whole rig very nearly tipping over. How it never did, I'll never know. There were an awful lot of grinding and creaking noises as the rig settled.'

Luckily for Pearce and Frost, Graham Wainwright and Bob Paul were following in another rig and were able to help. Wainwright added:

'I could see Gordon's truck go into the ditch so Bob and I pulled up alongside and quickly attached a long chain from the side of our truck up to the top of the container on Gordon's. We hoped it would stop the whole thing from tipping right

Chains were attached to the sunken rig to stop it from falling any further while they waited for more help. (GW)

A distraction caused Bob Paul to let the rig run into a ditch. Winter 1969, Eastern Turkey. (BV)

over. Bun and Eddie Parlane were also in the convoy and we attached another chain from their truck as well.

'It was touch and go with all of us praying that the chains would hold and stop Gordon's rig from toppling right over. As it was dark, we had to wait until morning to assess the situation, so we just sat there staring at the rig. None of us could believe it was still upright.'

When daylight broke, it became apparent exactly how close the whole rig really was to catastrophe. As the men were discussing what to do next, an Iranian truck driver stopped to offer his help. Wainwright continued: 'We attached yet another chain from the Iranian truck to Gordon's front bumper and told the driver to wait until we were all ready and then to move off very carefully.'

When everyone was set, the signal was given and all the trucks moved at the same time. With an awful lot of creaking and groaning, Pearce managed to steer his truck out of trouble and back onto terra firma. Pearce couldn't believe that his brand-new Scania rig came out of the ditch unscathed but, more to the point, the heavy transformer didn't break away from its restraints and fall off the trailer:

'I was very lucky that day and had it not been for the quick thinking of my workmates, I may well have lost the complete rig, the load and my job.'

Bob Paul shared trucks with his drivers right from day one and even though he was the boss, he always did his fair share at the wheel. 'My boys were all excellent drivers,' he said. 'But it sometimes wasn't their fault and they were powerless to do anything to avoid an incident. I have to admit, that included me too.'

Paul travelled with Bobby Vallas on many trips to Tehran and recalled one favourite story:

'I was driving with Bobby asleep on the bunk. It was snowing and we were near Agri in Eastern Turkey. It all happened so quickly. One minute I was on the road, the next I was in the ditch. I must have drifted too close to the edge and got dragged off into the softer ground.

'Vallas woke with a start and jumped out of the cab to see what the damage was. He came back and said in his broad French accent, "Bob, we'z in ze sheet. Ze axle az gone."

'The rear trailing axle had been badly damaged and there was nothing we could do ourselves, but as we were in Turkey I was confident that we could get it repaired. It cost us a week, but we managed to find some local help to get the truck towed out of the ditch. I found a blacksmith to weld up the axle which got us back on the road again.

'A little further on, the weather was even worse. I was driving again and Vallas was sleeping … again. I was on

The pins on the A-frame drawbar coupling came away causing the trailer to break free from the truck and roll over. (RS)

snow and sheet ice by now and noticed a local Tonka coming up the slope towards me. I could see that he didn't have any snow chains fitted and I knew instinctively that he wouldn't move over from the centre of the road. So it was I who had to yield and once again I let myself go into the soft snowy verge to avoid colliding with him.

'Our truck ended up at a very bad angle embedded against the hillside. Again, Bobby woke with a start and jumped out to assess the damage. I remember him getting back into the cab and saying, "Bob, ziz time we'z in ze fucking sheet."

'We were miles from anywhere so I had to walk to look for help. I came across a military camp where, although no one spoke any English, the commanding officer did speak French so at least we could communicate a little. He rounded up some of his men and they brought machines and even horses to help pull us out of the ditch.

'Luckily there was no mechanical or structural damage to the rig so after we had sorted everything out, Bobby and I got pissed on raki with the CO. It was very satisfying.'

On a return trip from Iran in the late 1960s, Ray Scutts was sharing a Scania Vabis with John Frost who was driving while Scutts was getting some sleep. Scutts takes up the story:

'When Frosty approached Bazargan he gave me a shout to wake up. As I got my bearings, I looked outside and got a huge shock as I saw what I thought was another Asian Transport truck about to overtake us. I said to Frosty, "Who's this prat coming past?"

'Then I suddenly realised it was our trailer. I couldn't believe what I saw. It came alongside us, then veered off and barrel-rolled to the side of the road. Of course we stopped as quickly as we could and found that the towing hitch had come clean away from the front of the trailer.

'It was the rough roads which had probably shaken the pins out and even though we made regular checks on everything around the rig, there was always the danger that something like that would happen.

'We found some of the hitch parts scattered all over the road but it was the trailer that was so badly damaged. The roof had been completely ripped off and the precious load of Persian carpets was all over the place. We had about a million pounds' worth on board.'

Scutts and Frost had to leave the truck and wrecked

Expensive Persian carpets worth a cool million pounds were scattered all around! (RS)

trailer where they were and go to the customs office nearby to explain exactly what had happened and also to advise their office back in England. Scutts arranged for an agent to make out new documents so that the men could transfer the trailer load to another vehicle. Scutts continued:

'We took the truck to the border and emptied it into the customs warehouse. Then we went back for the carpets from the trailer. A day or so later we hitched up to a hired trailer that our office had arranged for us and we were able to reload everything and continue homeward.'

The damage was so great to the original Asian Transport trailer that it was beyond repair and so it was left exactly where it fell. It was soon vandalised and Scutts remembers often seeing it as he went to and from Iran:

'It was there for about two years and became a bit of an icon. The stories soon got round about it, too. People used to say, "Have you seen the trailer that Asian lost in the mountains where it got shot at?" It was all quite funny really.'

Scutts also had a problem or two when he took Dick Snow out on his very first trip. Again Scutts recalls the story:

'We had got stuck on a Turkish mountainside and had to find the council workers with their bulldozers to tow us out of the snow drift. When they arrived they didn't hang about and hooked a chain onto the bumper of our truck. Then before either of us could jump in the cab, another Turk shouted to the dozer driver to move and he just took off, almost pulling the bumper clean away because the handbrake was still engaged.

'I'll never forget that because the truck was brand new and on its first trip. It looked a right mess and we had to put up with it until we got home. It didn't do our reputation as drivers any good at all.'

During his driving days Peter Cannon had a few mishaps. The most memorable one had to be an incident in Austria while he was driving a Scania 110 road train, AMY 147H, accompanied by Tony Soameson. Cannon takes up the story:

'We had stopped at the border into Germany for the weekend so decided to take the rig into Salzburg for something to do. We got a bit lost trying to follow the signs

WLO 95G on its inaugural trip. Turkish council workers pulled the bumper off with a bulldozer. (GP)

and unfortunately ended up right in the centre of the city which wasn't really the place for big trucks.

'I was driving and noticed the tram wires hanging just above the truck roof. I couldn't turn the rig around, so I drove very slowly under the wires which, luckily for us, didn't touch the roof of the truck so I continued underneath them.

'Unfortunately the trailer was a bit higher than the truck. As the trailer went under, the wires caught on its roof causing sparks to rain down everywhere. It was like a fireworks display.

'Before either of us could do anything, a fire started and very quickly the thin aluminium trailer roof began to melt and fall into the load which in turn caught fire. I was so worried. The fire brigade was called and eventually they put the fire out.'

The local newspaper ran the story the next day which, in translation, reads:

'It has already been reported yesterday that a British lorry driver has caused a lot of interest in the city. On Wednesday, the trailer cut through the tram lines which broke and came down on the trailer roof. After the fire brigade arrived, they extinguished the fire in a short time.

'The truck was carrying carpets and furs. The estimated damage was around three million schillings. The cable damage was around 25,000 schillings. Due to the accident the trams were out of service for 4½ hours.'

As well as his responsibility as transport manager, Peter Cannon was also tasked with investigating incidents and accidents involving the trucks and drivers. It was not unusual for him to jump on a plane at a moment's notice and fly off somewhere to sort things out. That suited Cannon as he could use his own initiative to solve the problems. As he said:

'Most of us today have constrictions in our working life. We have laws that say we can only drive nine hours a day or that is the route you must take, or you must telephone the office if you have a problem with the truck.

'In the olden days, we had to think for ourselves. I couldn't just phone Bob Paul and ask him what to do because he didn't know, either. I was on my own and I had to do what I thought was best at the time. I enjoyed the responsibility and the challenge of getting everything sorted.'

Cannon's calm nature stood him in good stead as he had to deal with many irate officials during his investigations:

'You had to be realistic and know how things worked in different countries. I had to learn who to bribe and who not to bribe because in some countries that was how things worked.'

Cannon would receive a telegram in the office detailing what had happened and where. He then packed his bags and made for the airport. Here he tells the story of one such incident which he thought would be quite simple to sort out, but in fact turned out to be far more complicated than he could ever have imagined:

As the trailer roof hit the live tram wires, all power to Salzburg's trams was cut instantly! (TS)

'Shortly after I came off the road in the early 1970s to work in the office, one of the trucks, AMY 147H, was involved in an accident in Bulgaria when it was en route to Iran. All we knew was that the truck had tipped onto its side and driver Ronnie Pinkerton was in hospital. My tasks were to help the driver, sort out the load and the damaged rig, and then continue the trip to Iran in it.

'When I got to the town where it had happened I went straight to the hospital, only to find Pinkerton had discharged himself. One of the nurses suggested he might have gone to a local hotel so off I went in a taxi and sure enough I found him with a couple of broken ribs, cuts and bruises. It was obvious that he couldn't drive the truck so I arranged an air ticket home and got him to the airport.

'The truck was still on its side and among the cargo was some whisky which was leaking. I contacted the police who had already been to the scene and asked them to arrange some labourers and I also asked if they had such a thing as a bulldozer. I got a local farmer to supply some straw bales to put round the truck and I managed to get some people to sit and guard it all night so that no one stole anything.

'The next day, with help from the labourers we all unloaded just about everything and piled it up on the side of the road. Then we attached some chains to the containers and used the bulldozer to pull the truck carefully out of the ditch. Then of course we had to load it all up again.

'While this was happening, the customs officials were always present to make sure that none of the precious cargo went astray as it was all under TIR regulations and in theory shouldn't have been touched. I had the difficult task of making out a protocol, a long list detailing which goods were in the truck and which were in the trailer and also what had actually happened.

'The customs made a copy of the protocol and also wanted to know why some of the goods that were originally on the truck were now reloaded on the trailer. It was a nightmare to sort out. If it didn't all tally up, I could have been accused of illegally importing goods and then selling them.

'Once the load was sorted, there was the matter of a big hole in the trailer roof to patch up and the drawbar coupling which had been ripped off and needed welding back on. I managed to find a local garage and a blacksmith to come and make some temporary repairs. Just to add insult to injury, there were a couple of hoses burst in the engine which lost a lot of oil and water – and all this took a great deal of time to get fixed.

'Eventually I was able to drive the truck and got it through the border at Kapikule and into the Turkish town of Edirne. Unfortunately, I couldn't find anyone there who could weld aluminium so I had to continue to Istanbul where I had more luck. I left the truck with the welder and stayed at a hotel for the night.

'In the morning a waiter approached me. He said that three men sitting at a nearby table were secret policemen and they were talking about me. Apparently, the waiter had overheard them saying that I had a customs TIR seal missing on my trailer. I did a bit of quick thinking because I didn't want the policemen to know that I'd been told.

'I went to the truck and started walking round checking the tyres, lights and so on. Then I noticed that one of the seals was indeed missing. I wasn't sure if the police were watching me so I had to look surprised. I decided the best thing to do was to take the truck to a local agent and get the trailer re-sealed.

'I couldn't park outside the agent's office so I left the truck in a car park and started walking back up the road. All of a sudden I felt someone grab me and turn me round. It was one of the policemen from the hotel. In perfect English he asked me where I was going. I told him what had happened and where I was going. He replied that he and his colleagues would accompany me.

'When we got to the agent, there was a message for me to telephone Bob Paul immediately. The outcome was that my father had died so I decided there and then that I had to fly back home. The agent told me that I couldn't do that because the police thought I had imported guns into Turkey, broken the seal on the trailer and sold the guns. He said that because the heat was on, they thought I'd made up the story about my father so I could get away.

'The agent advised me to act normal and drive the truck through Turkey. When I protested he responded, "What would you rather do? Be late for your father's funeral or spend ten years in a Turkish jail?"

'The onus was on me to prove that I hadn't brought any guns in or sold them. I showed the policemen the protocol which the Bulgarian customs had stamped and signed but the police couldn't read it because it was all in Bulgarian Cyrillic script. I suggested that they got it translated but they weren't interested. Instead they insisted that I go to the customs compound with them and have the complete load checked against the paperwork, which is what I had to do.

'There was another lengthy delay and to make matters even worse, the officials couldn't get everything back in the way we had done it in Bulgaria. The load had now been rearranged twice which really did cause some problems with the documentation. Eventually, after much debating and arm waving, the Turkish officials seemed happy with things. As luck had it, another Astran truck had arrived driven by Ron Bell and Ray Neal so at least I had some moral support.

'The agent told me that the best thing was still to carry on acting normal and drive the truck on through Turkey as the policemen still had their suspicions about me. So the three of us drove the two trucks non-stop, a thousand miles across the remainder of Turkey.

'I knew there was a British Airways flight from Istanbul to England at 4 pm every day, so I thought if I could get to the far end of Turkey at Erzurum in time, I might be able to get an internal flight back to Istanbul and then pick up the BA afternoon flight home.

'All went well until we got to the Turkish exit border at Gürbulak. I said to the others, "Right, that's it. I'm off," but the customs man wouldn't allow me to go. Every single detail about me, the truck, the load and the document numbers were all written in my passport and I was not allowed to leave the country unless I took everything with me. That was their rule.

'The truck was not allowed to leave either unless I went with it. That was another rule. It also transpired that neither Ron nor Ray could take it because my details were logged against the truck in my passport and only I could drive it. I was stuck. There was only thing I could do and that was to continue driving the truck towards its destination of Tehran.

'With Ron and Ray following, we crossed the border which, ironically, I had no trouble doing. Once I got into Iran the officials were not interested in my previous business with the Turkish authorities. I now had no chance of getting a flight back to Istanbul and so I telephoned the agent in Tehran and asked him to get me a ticket to London from the airport there. It was early morning and I explained to him that my father's funeral was the next day and I really needed to be there.

'He agreed to that, so I left the truck for one of the others to drive and jumped in a taxi to Marand, the first town in Iran. When I arrived I found a bus company but the journey to Tehran was still 550 miles and they had no buses leaving for another four hours. I was very conscious of the time and the taxi driver knew this, so he kindly took me to the next bus company. It was the same situation there with no buses leaving. Eventually we found a company that had had a bus leave an hour beforehand. The taxi driver was not at all deterred and told me he could catch it up.

'I must admit it was a bit frightening as we roared along the dusty roads at 100 mph but, give him his due, he knew where to go and we eventually caught the bus up. It was like something out of a film; I threw some money at the taxi driver, jumped onto the bus and crammed myself into a seat with my knees jammed under my chin.

'I was dirty and extremely tired but still eager not to miss the funeral. When we finally got to Tehran I spotted the huge monument which I remembered was near to the airport, so I jumped up and shouted to the bus driver to stop and let me off. I hailed another taxi and made it to the BA check-in desk.

'Literally breathless and sweating, I asked the lady if she had a ticket for me. "Oh yes, Mr Cannon," she replied. I was so relieved and thanked her profusely.

'I asked her where the plane was. "Oh, it has already left," she replied. Now I was flabbergasted. It was the day of the funeral and I was stuck again.

'I explained to the clerk why I had to get home so urgently and everyone was so helpful. They made a couple of telephone calls and actually managed to stop the plane before it took off. It was a jumbo and I really couldn't believe what these people had done to help me. They even sent a Land Rover to collect me from Departures and take me out onto the runway. I felt like a proper VIP as I climbed the steps. I then finally managed to relax and settle down as we took off.

'Next thing I knew, there was an announcement: "We are approaching Tel Aviv airport." It dawned on me that my plane was a round-the-world jumbo which did exactly that. It flew all around the world stopping at every major airport. It was still very early in the morning and I worked out that it would be another seven hours until we landed at Heathrow. I would just about make the funeral in Norfolk if I had all the luck in the world on my side.

'We touched down at lunchtime and I got a taxi to Liverpool Street train station by 2 pm. The funeral was at about 3 pm and I now knew that I had no chance of making it. I got a train to Kings Lynn and then another taxi and I met the family for the evening reception. I was totally shattered but at least I got there in time for a sandwich and a beer.'

This is a good example of a job which on paper looked like quite a simple task but turned out to be fraught with delays, confusion and even danger. Cannon said that he feared for his safety and at one point thought he might be thrown into jail, never to be seen again.

Another driver whom Cannon had to rescue was Kim Belcher who was unable to control his truck, WHJ 194S, down a hill in Turkey. Belcher ran off the road and suffered serious injuries to his arm.

The truck was too badly damaged to be driven so Cannon had the unenviable task of changing all the relevant documentation plus Belcher's passport, to allow the truck to be separated from its driver. Cannon met the British consul in Ankara as it was he who had to write an official letter to the Turkish authorities explaining the situation and asking them to adjust the documents. Once everything had been signed, sealed and stamped, it was a simple case of arranging another Astran truck to recover Belcher's unit and trailer to the UK.

Belcher still has a nasty scar on his arm to remind him. 'The accident very nearly put me off driving trucks forever,' he said. 'But that Scania was so well built, it saved my life.' This was how the accident occurred:

'I was coming out of Ankara on the way home fully loaded with personal effects from Doha. I got to the top of the notorious Death Valley on the way to Istanbul and I noticed that it had been raining. One side of the road was bone dry but on the other side it was still wet. It was dark and I couldn't really see where the rainwater was. As I hit the brakes half-way down the hill, the truck just started swaying right to left and I couldn't stop it.

'I was petrified and before I knew it, the whole rig just tipped over onto its side. It ended up in a ditch and on the only bit of bank next to the road. I was so lucky it hadn't gone on a bit further because I would have probably died if the truck had gone over the edge into the ravine!

'My left arm went through the quarter-light window and severed the nerve to my wrist and I had some glass in my eye. I was ejected from the cab through the back corner window above the top bunk but my knee got caught under the steering wheel so I ended up with my head and shoulders out of the cab looking straight at the huge diesel tank which was still full. The batteries were right by my head and I thought, "Oh shit, I'm in serious trouble here!" It freaked me out a lot.

'Before long some Turks arrived to help. They attached a chain to their Tonka, then the other end was attached to my cab and they tried to pull the whole thing back up so they could crawl underneath and get me out. Every time the Turk put his handbrake on, the whole lot just slid back down the bank due to the weight in my trailer.

'Eventually someone put the little Tonka into crawler gear and just kept spinning the wheels to hold it all while two others

Kim Belcher couldn't control WHJ 194S as it came off the bend. (KB)

Belcher was very lucky to escape. Had his truck gone further down the hill, he might not have been so fortunate! (PC)

dived underneath and dragged me out. It was the bravest thing. They were fantastic.

'The army turned up later and put the truck under guard while I was taken to hospital. I managed to go back to the truck the next day because I wanted to find out who the men were who dragged me free - but the army captain wouldn't tell me.

'The biggest problem I had was the Turkish hospital services. I was literally dragged to Ankara where they stitched me up and shot me full of morphine. It was about thirty-six hours before I realised that all my teeth were caked in mud where I'd been dragged along the ditch.

'The British Embassy also got involved. Peter Cannon flew out to help with the paperwork, especially the difficult task of separating me and my truck from my passport. I flew back to the UK where I was operated on in hospital to repair my damaged arm. After I had convalesced for a couple of months Bob Paul asked if I would be okay to drive again, so I gave it a go. However, my bottle had completely gone and it took a long time to get over it all.'

Surprisingly, the truck that Belcher was driving, WHJ 194S, was not too badly damaged and was recovered to Ankara. Then it was piggy-backed home to Addington where it only needed a new cab and some cosmetic work to make it roadworthy again.

There is some confusion as to who was actually driving another Scania which careered down a bank in the same area as Belcher's accident. There were two Astran trucks running together. One was driven by Ray Thorne and the other by a friend of his, name unknown, who was on his first trip for Astran. According to Barrie Barnes, Thorne's truck had around three tons of freight on while the other had twenty-three tons.

Barnes believed that, as they were both descending the long hill, it was Thorne's mate who got brake fade and simply ran out of brakes. As he was constantly pumping the brake pedal, it heated the parts up so much that they became red hot and totally ineffective.

As the truck ran off the road, it took a local Tonka with it and then hit another one at the same time. The trailer was loaded with pipes but amazingly none of them moved when the rig went over the edge. The unit jack-knifed, but luckily the driver only suffered slight injuries to one of his legs where he got it trapped under the steering wheel. After a brief spell in hospital he was released and spent the next few days in a local hotel while his truck was recovered and his documents and passport were dealt with.

Barrie Barnes spent a month in Ankara waiting to get the load sorted and cleared by the authorities so that he could deliver it on his trailer. One of the jobs Barnes had to do while he was waiting was to hire a machine to dig out the bank and build an access road so that the crashed Astran rig could be recovered and returned to Addington for repairs.

Being in the wrong place at the wrong time was definitely a hazard of the job. John Williams had his fair share of troubles and was interviewed in the Dubai publication, *Khaleej Times*, about some of his incidents. Here is his account of one of them:

'I had seen in a Turkish newspaper a photograph of a Turkish driver lynched by a mob after he had run over someone, then not long after that, I knocked down an old man in Iran! He ran out from behind a bus and I had no chance of stopping.

'Within five minutes, there must have been two hundred Iranians there. Seventy-five per cent came to see what was going on, five per cent were intent on hanging me on the spot and the rest were trying to stop them. Two hours went by, no police turned up and the chap looked like he was dying.

'Well, I didn't think I could control the situation any longer so I bribed a bus driver to take him to hospital. When he went, the crowd vanished too. During the two hours I had been there, I stopped several vehicles and asked them to get the police, but after another five hours they still hadn't arrived so I moved my rig which was in a dangerous position.

'Immediately I did that, the police drove up and the first thing they said was, "You're a very bad man for moving the truck." I explained why I'd done it and they said, "Okay, you're only half-bad. You stay, the gendarmes are coming."

'By that time I was firmly convinced I was going to be locked up forever, so I stuffed my sleeping bag full of cigarettes, tins of food, soap and toothpaste. I waited for another three hours and in the end I went to bed.

'About three in the morning there was a great Bang! Bang! on the cab and there stood two gendarmes. They motioned me to get out so I took my sleeping bag and we started walking up the road. I asked where we were going and they said a town a hundred kilometres away. "On foot?" I asked.

'One of them replied, "We'll get there tomorrow or the day after." So I stopped and said we could go in the truck if they liked. They had a little think about it and agreed. When we got there we went to a prison and they knocked up the commandant.

NVW 480P crashed down the bank but amazingly none of the load in the trailer moved. (DP)

The Astran rig took one Tonka down the bank with it and hit another on the way. (DP)

Slips, Trips and Falls

Barrie Barnes, driving WHJ 193S, pushes the stricken NVW 480P onto a stripped-out trailer so it could be recovered back to Addington. (BB)

'"Is he a good man?" he asked the gendarmes. "Well, he can't be all that bad," they replied. "He waited eight hours to get arrested."

'Then the commandant turned to me and asked, "Are you a good man or a bad man?" "Oh, I'm a very good man," I replied.

'He pointed to a cell and said, "You sleep in there or you sleep in your truck." I replied that I would sleep in my machine, so he said, "Six o'clock in the morning you come here."

'At six when I hammered on the great prison door, they grabbed hold of me, threw me in the cell and locked me up. Then at six at night they threw me out again. This went on for seven days! Eventually I reached an agreement to pay the injured man's family and they released me.'

In the early 1970s subbie John Cooper came across Ron 'Jock' Bell and Noel Tehan who were paired up in a Scania Vabis en route to Iran. They were having difficulties negotiating the Tahir pass. Cooper explained:

'There was a very steep hill which had a small bypass going around it that wasn't so steep, but it was extremely boggy. Jock chose to take the bypass option but half-way round he got the complete truck and trailer bogged down. He decided to drop the trailer and then managed to get the truck free.

'He then drove for a mile or so until he could turn round and come back to the trailer as he thought he could pull it out using the tow hitch on the front bumper of the truck. Unfortunately poor old Jock couldn't get proper traction and the wheels just kept spinning He tried to push the trailer backwards instead, but just ended up getting into a situation worse than he was in to begin with. He finished up with the truck and trailer facing each other, both completely bogged down.

'It was obvious that a big, powerful truck was needed to extricate everything and as luck had it, Bobby Vallas came along in his Scania, TDX 400K, which had double-drive axles. He dropped his trailer and used his tractor unit to pull the Vabis and trailer out of the mud in one go. Vallas made it look so easy.'

Like most of the drivers, Terry Tott was not a fan of driving anywhere in the harsh winter months:

'The main trouble in the winter was finding somewhere safe to park, especially if it was bad weather. I wanted to get through the harsh conditions so I'd been driving four or five days without much proper rest and it had already taken me about a day and a half to get to the summit of a particular mountain in Turkey. When I got up there the wind was horrendous.

'I had a mate with me and when we stopped the rig, he commented that he thought we were still moving. He wasn't wrong. The road was covered in snow which had been compacted and turned to sheer ice. Even though I had snow chains on the rig, the very strong wind literally pushed me off the road into the ditch. There was nothing I could do as we slid over.

Ron Bell tried to extricate his trailer from a muddy track by using his truck, UHM 26F, to push it backwards. Unfortunately, Bell got the truck stuck too. (JC)

'I just clung onto the steering wheel. We sat there for a while wondering how we would get out. As luck would have it, Colin escaped from the cab just as the whole rig tipped further onto its side! I was okay and clambered out too.

'As we stood there looking at the stricken rig, a Turk pulled up and offered to tow us out. I remember his comments clearly. He said, "Scania no good. My Volvo will pull you out." We hooked a chain to the front bumpers of our trucks and away he went in reverse. All he did was pull the bumper clean off his truck!

'Then he changed his mind and said, "Scania strong, Volvo no good."

'It was obvious my truck wasn't going anywhere so the Turk took us down to the village where we contacted Turkish customs to explain what had happened. The only way to get my truck out was to unload it on the roadside which meant breaking the sacred TIR seals. It was quite a palaver to arrange.

'The army was despatched to guard the truck and its valuable cargo of carpets and I managed to find someone who had a huge bulldozer. We also had some local labourers who came to unload the trailer, all of which was done under the watchful eye of a customs man. Then we managed to drag the empty trailer backwards off the tractor unit with the bulldozer.

'We got it back onto the road and then dragged the unit out in the same manner. Eventually we loaded the carpets back onto the trailer and got it re-sealed with the customs officials present. I think the whole episode cost about £200 which was a lot of money in those days.'

It had taken Tott a good couple of days to recover his rig and get back on the road. The authorities had only allowed him a certain transit time to drive through Turkey and the delay had pushed him over that limit. When he got to the Turkish/Bulgarian border at Kapikule, he found himself at the end of a long queue.

Tott went to an agent for help and was amazed when he was escorted right to the front and processed in double-quick time. 'The incident did me a favour really,' Tott said. 'That was the fastest I've ever been through any border.'

Tott telephoned Bob Paul to tell him about the accident:

'The incident happened two days after New Year. I spoke to Bob at his house and told him I couldn't drive the truck any more. He asked why and I told him I'd broken a mirror glass. He told me not to be stupid and to buy another one – to which I replied that I couldn't because the truck was lying on it. He went very quiet on the other end of the telephone.'

Tott was proud that this was the only incident he personally had while working for Astran, although he did come across other colleagues who were in trouble.

Andrew Wilson Young was one such person, as Tott described:

'I was on my way out to Iran and spotted Andrew's rig lying in a ditch in Eastern Turkey as he was returning to the UK. I think he must have lost control as he came round the bend and down the hill. I stopped to help and at first I couldn't see him so I banged on the cab and found him fast asleep in bed. Typically, he jumped out wearing a pair of shorts and his famous wellington boots.

Terry Tott (pictured) was unable to stop SOO 815R from being blown off the slippery road by extremely high winds. (TT)

'His rig was stranded and he needed my help, so I turned my truck around and fixed a chain between us. It was useless as my truck just kept wheel-spinning and wasn't powerful enough alone. Luckily there was another truck nearby. We chained that one to me so we had two trucks to pull him clear.

'It was a long job, but eventually we did it. We were all very cold and somewhat exhausted and I remember asking Andrew why on earth he hadn't got his snow chains on to which he replied in his distinctive posh voice, "I didn't think I needed them, old chap." I was hopping mad!'

On another trip in later years Wilson Young was travelling quite fast through Romania when all of a sudden the road narrowed. Young's truck clipped a bridge parapet which severely damaged the cab and twisted the chassis.

On that occasion Dave Poulton was the driver who went to his rescue:

Andrew Wilson Young became stuck in Eastern Turkey. Terry Tott stopped to help and was annoyed to find that he hadn't used snow chains. (TT)

Wilson Young's trailer was too high for German roads, so a special permit was needed to run it through the country. (DP)

'I was coming home empty and was going to load somewhere in Europe but I got a message on the way through Turkey telling me to go and help Andrew. I found him and his truck and then we spent all day loading it onto my trailer. A huge Russian truck with a crane fixed to it was already on the scene and together we gingerly lifted Andrew's rig onto mine in one go.

'I wondered if the crane might cost us an awful lot of money and was prepared for the worst. After we completed the loading, the crane operators simply settled for a carton of fags and a bottle of drink. I was happy with that, I can tell you.'

Poulton was concerned that the height of Wilson Young's rig on the back of his trailer would be too high to travel and indeed he was correct. It wouldn't fit under the telephone wires along the roads, so a stop at a local welding shop was called for. The front of Wilson Young's trailer was cut off to make it the same height as the tractor unit cab and then the journey home could begin properly.

However, Poulton was still not convinced about the overall height:

'We got to Germany and the officials wouldn't allow us to enter the country because they said the rig was too high for their bridges. They even measured from the ground to the top of Andrews's trailer and told me it was just a couple of inches too high. I knew it wasn't but that was typical German bureaucracy.

'We had to sit and wait at the border until a special travelling permit was brought to us to allow us to continue through the country. It all took a ridiculous amount of time which impressed neither Andrew nor me.'

During the early 1980s, Alan Warner was driving a new Scania 142 on a trip to Doha, loaded with crates of machinery. Unfortunately Warner lost control of his vehicle south of Baghdad and the complete rig rolled down a bank causing both tractor unit and trailer to be badly damaged. The load, which was packed inside eight-foot-square crates, was also damaged beyond recognition and was written off by the insurance company.

John Bruce wasn't far behind Warner when the accident happened and helped his colleague as best he could. While Warner was taken to hospital for some treatment, Bruce arranged for a crane to come to the accident site. A third Astran driver, Denis McGrath, was also in the area and while Bruce and McGrath were clearing up the mess, word quickly got round the local community that a truck had rolled over nearby.

Before long a local man appeared to offer his assistance and kindly stayed with the men through the

In October 1983 Alan Warner was driving this new Scania when he lost control in Iraq. (JB)

night to act as a guard while they got some sleep. After the cargo was removed from the crates, as arranged by the insurance company, the local man asked Bruce if he could have one of them. As a gesture of thanks for his help, Bruce allowed him to take a crate. Bruce clearly remembers the sight of the small Iraqi man trying to roll the heavy wooden crate up and over the hill:

'He started rolling it at about 6.30 in the morning and we could still see him at 5 pm that same day. He was still in our sight as he rolled the crate very slowly out into the desert and over the horizon. We couldn't believe what he was doing but that wasn't the end of it. About a day and a half later, he came back for another one. It was amazing.'

Barrie Barnes was also involved in a nasty accident while driving a new truck in Turkey when he ran into the back of a Tonka which, surprise, surprise, was grossly overloaded with peas and cucumbers. Barnes said that the impact was so bad that it knocked the cab of his Mercedes clean off the chassis.

The truck was only on its second trip to the Middle East and Bob Paul was unimpressed when Barnes informed him! Barnes went to a local police station after the accident where he was questioned and breathalysed. 'That consisted of breathing onto the face of the local doctor,' he chuckled.

The Turkish driver whom Barnes hit was also taken into custody and after much deliberation between all parties, Barnes gave the other driver £1,000 for the loss of his truck and its load. Kim Belcher was also in the area at the time and was able to help Barnes by piggy-backing his damaged truck and trailer back to the UK.

The late Billy Russell was one of Astran's regular sub-contractors on the Middle East run. On one trip home he had the misfortune to roll his rig near Adana in southern Turkey. Unfortunately he suffered bruises and a broken shoulder. Astran arranged for subbie Tony Soameson to divert and collect him which made for yet another piggy-back load to the UK.

Russell and Soameson had different notions of tidiness and the enforced close habitation in Soameson's cab during the homeward journey proved too much for both of them. Russell was lackadaisical whereas Soameson was neat and tidy. By the second night they had fallen out and Russell was consigned to his own cab, crushed and without windscreen, for the rest of the journey!

The cab of Barnes's new truck was knocked clean off of the chassis! The orange indicator should be in line with the centre of the wheel. (BB)

NKM 284W returned at Addington where it was forklifted off the trailer. (BB)

Slips, Trips and Falls

Mike Taylor was given the job of collecting Russell's wrecked truck from the docks and taking it and Billy back to Pembrokeshire. Taylor arrived at Ramsgate to find a terse note pinned to the front of Soameson's trailer: 'HE is in his cab.'

Not all the accidents were caused by other vehicles as Rick Ellis found out:

'Mike Walker and I were in two trucks heading for Muscat and were pushing on through Saudi Arabia, having left the Ar'ar border from Iraq earlier that morning. We were doing very well and had got to Nuayriyah just as the sun was going down, so we decided to stop and eat. We set off again that night, heading along the famous camel road to Hofuf with the intention of getting out of Saudi and sleeping on the Qatar side of the border.

'About an hour after being on the move again, I was tweaking my short-wave receiver and trying to connect with the BBC World Service when a camel came from nowhere and crashed into the front of my truck!

'In hindsight, I suppose that had I been concentrating on driving, rather than fiddling with the radio, I might have seen it coming even though it was night time. The headlights may have picked the beast out in the darkness.'

For those who may not know, a full-grown camel carries around 250 kg of 'motions' in its intestines at any one time. On impact Ellis's camel discharged most of its own motions all over his truck!

Under Saudi Arabian law, if anyone hits a camel in daylight hours, it is their fault. However, if anyone hits one at night, then it is the camel owner's responsibility. Ellis and Walker went straight to the nearest police station and reported the incident. They also went back to the station the next morning to get the relevant documents processed. The owner of the camel was never found, 'but I am sure he would have been found had it happened in daylight hours,' Ellis said.

Mike Taylor had the job of taking Billy Russell and his wrecked truck back to Pembrokeshire. (MT)

The morning after the accident. It soon became very clear that the truck was undriveable. (RE)

Ellis made a call to his wife, Julie, whose first response was to ask how the camel was. Ellis responded: 'Fuck the camel. You should see the state of the truck. And it bloody well stinks to high heaven, too.'

Ellis's truck was in no fit state to go anywhere and he couldn't leave it on the side of the road, so he and Walker connected air and electrics between the rigs and fixed a towbar in place. Then Walker towed him through to Muscat. Ellis had a rough ride:

'Needless to say I had no engine so there was no air conditioning and no power steering for the next three days. It was terrible trying to steer the rig and the ride was unbearable.

'If that wasn't bad enough, just after we entered Oman at Wadi Jizzi, there were about forty kilometres of wadis – dried-up rivers – and Mike towed me through every one of them at 60 mph without stopping once. I was six feet behind him and it was utter hell! I clearly remember jumping out of my cab when we finally stopped and chasing him round the truck with threats of death.'

After both trailers were unloaded in Muscat, Walker put Ellis back on tow again and took him back at a more leisurely pace into Saudi. There a workman who had a Cat D8 bulldozer working in the desert made a ramp in the sand. With help from a Jordanian driver, Ellis's truck was pulled onto Walker's trailer in one go. The height left Ellis nervous:

'Talk about the backside twitching. We had a four-metre overhang and we knew we were going to be too high for the bridges across Europe. So we stopped in Turkey where we spent a fair amount of time taking the wheels off to reduce the overall travelling height.'

CHAPTER 15

Destination Doha

During the 1970s the BBC broadcast a highly successful television series called *The World About Us*. This weekly documentary was both inventive and adventurous. The objective was to bring sharply into focus many facets of life outside the experience of most people.

One of the men involved in the production department was Simon Normanton who came up with the idea of following a bunch of truck drivers as they drove to the Middle East:

'I was a researcher and we were looking at countries that people knew nothing about in an attempt to make unusual television documentaries. I was thirty years old and was always full of good and sometimes crazy ideas and was always dreaming up new programmes.

'I spoke to my boss, Tony Isaacs, and sold him the idea that a truck taking a load overland all the way to the Far East would be quite an adventure and it would be fascinating to try and record the journey. I knew Isaacs would like the idea because the madder it was, the more seriously he would take it. I had a potential winner.'

Normanton's project was prompted by an article which he had read in the *Sunday Telegraph Magazine* a year or so previously. In it, a reporter and photographer had accompanied two drivers from Asian Transport on a journey from England to Iran:

'I thought it was absolutely incredible and it made a huge impression on me. I mean, who on earth would want to drive a lorry some four thousand miles across two continents and then come all the way back again for goodness' sake!'

Isaacs gave the go ahead and told Normanton to make a reconnaissance trip so he could get a feeling of how the filming might go. So Normanton contacted Bob Paul:

'I told him what we were planning and asked if we could film his trucks. He was very interested and gave permission for me to accompany one of his drivers on a trip to Tehran, so off I went in 1976 with Roger Pierce. It was brilliant for me as I had a month on the road smoking King Edward cigars and getting paid for it.

'We went in the winter. I'd done a few crazy things in my life, but I'd never experienced anything like that before.

Simon Normanton accompanied Roger Pierce (pictured) in AMY 147H on a reconnaissance trip to Iran in 1976. (SN)

NVW 484P driven by Dave Poulton thunders across the desert, destination ... Doha. (SN)

Destination Doha

It gave me a fantastic insight into the unique lifestyle of a long-distance driver and how he survived the harsh conditions. I instinctively knew that Isaacs would love it.'

Isaacs was indeed impressed with Normanton's findings and decided to film the route to Iran, but to spice things up a little, he wanted to continue the filming to Kabul, capital of Afghanistan. Normanton made enquiries, but unfortunately the BBC could not obtain permission to film in either Iran or Afghanistan so another destination had to be chosen.

As Astran was also engaged on regular runs to the Gulf state of Qatar, it was decided to film trucks on that route instead:

'I secretly yearned to go to Afghanistan as the name alone made it all sound very mysterious and adventurous. Qatar didn't seem to have the same attraction and anyway a lot of people had never heard of the place.'

Normanton had also spoken to one or two other companies before he approached Astran as he wanted to see exactly who would be interested in the film. It seems that the others didn't see any potential in the project and thought it would be a waste of time and money. How wrong they were. The film that Normanton and his colleagues were about to make would eventually earn itself cult status as the best transport documentary ever.

In February 1977 all of Normanton's hard work came to fruition as the film crew arrived at Astran's headquarters in Kent ready to start work. For the next six weeks the five-man crew would be living in the pockets of four Astran drivers as they travelled on a 5,000-mile journey through eleven countries, from the tarmacadam of Europe to the sands of the Arabian Gulf. Their destination … Doha!

Normanton felt very excited but somewhat nervous as everything began to take shape. He'd spent many months preparing for the trip and at last all his efforts were about to pay off:

'We wanted the best film crew available, so I suggested Tony Salmon as the producer. I thought he would be the ideal person to head up the Astran film and Tony Isaacs chatted him up to come with us.'

At the time 34-year-old Salmon was working for the BBC on a motoring programme called *Wheelbase*, the forerunner of *Top Gear*. He was travelling all over the world filming various car rallies, races and safaris so he had the necessary experience to make the trucking documentary. Salmon was thrilled that Isaacs thought him the right person to head up the team, though in some respects this would be different from the work he had done to date:

'One difference with this film was that whereas on the rallies we'd go onto each location by air, here we had to go with the trucks all the way and it would be at a much slower pace. It would certainly take some getting used to and would be a huge challenge for me.

'It was a vital break for me because the film got me the calling card and allowed me to jump up to the next level in television. I was used to making 20- or 30-minute car programmes and now I had to make a two-hour film documentary. It was a very important film for my career and of course working on such a difficult subject also tested my skills to the limit.

'Foreign filming was no big deal, nor was moving along the road and working under different climates and in harsh conditions which were what I was used to, but this was much more specialised filming. Nevertheless, Isaacs put his faith in me as I had the expertise and knowhow to lead the crew.'

As Normanton was given the job of setting everything up, he was constantly in touch with Bob Paul at Astran. He also liaised with the embassies of the countries through which the trucks and film crew would be travelling, to make sure it would be okay to film at the borders.

As the film was being made by a British company and as Astran was transporting mainly British-made goods, it was also decided to try and include a British-built vehicle to promote the country's exports further. Contact was made with British Leyland who agreed to loan a newly built Leyland Marathon tractor unit for the duration of the filming, much to Michael Woodman's delight.

Arrangements also had to be made for minders to travel with the crew in certain countries. Salmon recalls that some of the minders were adamant about what could and could not be filmed:

'I'm sure they were instructed by their authorities to keep a strict eye on us because some of them would stop us filming in certain areas. I recall that to begin with, our Bulgarian minder never spoke any English which caused us a big headache when we wanted specific information. He would just blank us, then I remember saying something about his country and he suddenly answered in perfect English. I was furious.

Two Range Rovers used by the film crew were perfect for the arduous conditions, especially for tracking the trucks across the desert. The convoy at rest in Yugoslavia. (SN)

'Nevertheless, on the whole the minders were a godsend and we would've been lost without them. We also had vital help in Syria, Jordan and Saudi Arabia.'

For Normanton the hardest part of his research was trying to decide how the film crew would actually follow and film the convoy of trucks:

'We needed a special sort of rugged vehicle which we could equip specially for the filming. As I was already in touch with British Leyland I asked them for help. They came up trumps and supplied us with two Range Rovers fitted with heavy-duty tyres, roof racks, grab bars and huge driving lamps.

'We picked the vehicles up in central London and I remember thinking how good it was going to be, especially after we'd crammed all our equipment on board.

'I'd also stocked up on food supplies as I had no idea where we would end up at the end of each day. I bought huge quantities of dehydrated mashed potato, 'surprise' peas and catering packs of dehydrated chicken supreme and curry.'

The truck that British Leyland loaned was SLO 707R, a Leyland Marathon 'special' fully equipped for a trip to the Middle East. It even had a kitchen sink and, in keeping with the other trucks in the convoy, it was sprayed in Astran's livery specially for the film.

As the Marathon was a loaned vehicle, British Leyland also supplied a driver. On top of that, because the truck was brand new and a new design they supplied a third Range Rover – and driver – loaded with spares, just in case it broke down. There were grave reservations as to how reliable the truck would be compared to the Scanias in the convoy. The last thing anyone wanted was for the filming to be held up due to excessive breakdowns but, in the event, the first stoppage was for one of the Scanias.

Salmon said that he was very surprised indeed when his bosses gave permission to make a two-part programme:

'*The World About Us* was normally a single 50-minute programme and I am still amazed that we managed to get two 50-minute shows out of our filming because in those days that was unique.

'We shot everything on 16mm colour film which was very expensive. Colour television hadn't been around for too long either so many people were still moving over to it and that was another reason why the film was so expensive to use.

'A ten-minute section of filming would use up a 400 ft roll of film. It cost around £100 to have the roll processed in the studio so when we were shooting, not only did we have the editorial pressures such as which truck we should be filming, or which angle we should be filming at, but it was also very important to remember the money aspect.

'The film ratio was really critical when we were making the documentary. That was basically what we shot on location, compared to what went on the air. A modern drama programme can just about get away with a 4:1 ratio. Even with all the actors knowing exactly what to say and where to stand and with the cameras knowing every move, it is still very difficult to keep to such a tight ratio. So it was extremely difficult making Destination Doha because we simply didn't know what was going to happen next.

'We spent six weeks on the road so you can work out how many ten minutes there are in that amount of time. Our ratio should have been 8 or 10:1, but I think we ended up doing 12 or even 15:1 and so it didn't take many rolls of film before we'd bought ourselves an airline ticket each!

'We took boxes and boxes of film all packed into the Range Rovers. It was very heavy and took quite a bit of shifting at each location. We had two types of film: regular for daylight use and fast stock for the night-time shoots. Cameraman Alec Curtis used to get on to me all the time and hassle me to tell him how much daylight and night-time film I thought we would need for that particular day.

'I think we all did very well, learning as we went along, which was something we were all very proud of. But of course it wasn't just our crew

The Leyland Marathon 'special' which British Leyland loaned for the duration of the filming. (SN)

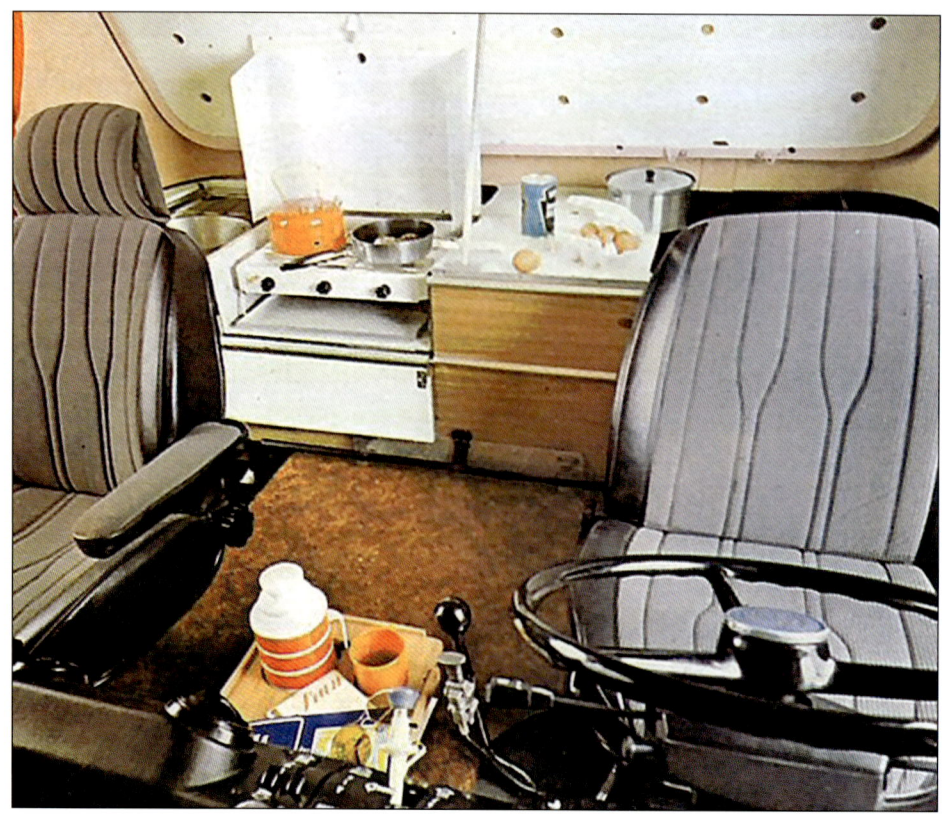

The interior of the Marathon was fully fitted and included a kitchen sink. (BL)

who should take the credit. Jim Latham also deserves immense praise. He was the most brilliant editor who put the film together once we had finished.'

And so to the film stars: Astran drivers Dave Poulton, Frank Hook and John Williams who were accompanied by British Leyland test driver Dick Rivers.

Williams was an excellent driver and an exceptional mechanic. He had good experience of the routes and was a natural leader so he was put in charge of the convoy and was seen throughout the film getting his hands dirty and making the all-important decisions. At the time of filming, Williams was going to Hofuf, Saudi Arabia, to repair and recover an abandoned Astran truck. Rather than fly there, he shared the cab with Frank Hook.

According to Hook, Williams's only bad habit throughout the journey was the fact that he was terrible at getting up in the mornings and refused to go anywhere until he'd had his first cup of tea.

At the beginning of the film, all the drivers are introduced to the viewer. In Frank Hook's case the narrator said that he was tasting the delights of the lorry marathon for the first time. On the contrary, Hook was not on his first trip to the Middle East. In

Although Dave Poulton was a star of the show he preferred to stay in his cab at night reading a good book. (SN)

Contrary to the film narration Frank Hook wasn't on his first trip, having already worked for Astran for two years. (SN)

John Williams (right) and Dick Rivers, the British Leyland test driver, debate which route to take. (SN)

fact he had been working for Astran since 1975 and had already made a significant number of trips.

Hook was somewhat disappointed with the way he was portrayed and only found out what was said about him when the film was aired on television later that year.

In fact the film makers had got Hook mixed up with Rivers, as it was he who had never been to Qatar before – although he had taken a truck on a test run to Iran for Astran in the early 1970s. He'd also included some European driving for other companies in his testing duties.

Like Williams, Dave Poulton was also very experienced on the Middle East run but always wanted to get things done as quickly as possible, sometimes quicker. Poulton was never one for socialising in the bars and would much prefer to stay in his cab at night reading a good book.

Dick Rivers was a thoughtful, unassuming man of whom the film narrator, Benny Green, said, 'If it's true that nice guys finish last, Dick will be bringing up the rear right through this trip.' Bob Paul did his best to ensure that Rivers was treated well:

'Dick was a lovely fellow and I was upset that British Leyland wasn't going to pay him properly for the trip. We were getting £5,000 for the load that Dick was going to deliver, so I told his employer they could have some of that money for the use of their vehicle, but they must pay Dick something out of it. All he was getting were basic expenses which weren't much at all, so I managed to get him quite a lot more. I remember him saying how grateful he was.'

To keep everything legal and above board the drivers had to become members of Equity, the actors' union, before any of them were even allowed to smile at the camera. That caused some whisperings at the Astran headquarters and the men were subjected to some good-natured ribbing by their colleagues for a long time afterwards.

Jan Payne in Astran's operations office typed all the documentation for the loads, the trucks and the film crew's equipment. However, it was transport manager Peter Cannon who was actually shown on the documentary handing all her hard work to the drivers when they departed. Payne was deeply disappointed:

'Even my husband had a part in the film if only for a few seconds. He was seen welding a truck in the garage, but unfortunately the film crew never came into our office.'

Normanton was apprehensive that the Doha trip might fall through. Astran was in financial difficulties at the time and right up until the trucks had actually boarded the ferry at Folkestone the whole project was in doubt:

'As film makers we had no inkling whatsoever as to what the problem was and we were relying solely on the drivers to tell us what was happening.

'There we were with all our expensive camera and sound

Simon Normanton had spent many months researching and now all his hard work was about to pay off. (ASTC)

equipment and the Range Rovers all packed up but nowhere to go. We were just sitting around waiting for the nod. I had an obligation to my employer to make the film and I thought we would look mighty stupid if nothing happened.

'Then as soon as we got onto the ferry I remember there were huge sighs of relief all round and celebrations by the drivers when we finally got on our way. When we disembarked in Belgium, there were even bigger cheers and we all agreed that at last the filming could begin in earnest.'

That relief could clearly be detected from the narration and from the comments made by the drivers in the film.

Normanton remains proud of the film:

'It is amazing and totally for real. I can't believe we shot it all with just one cameraman and one sound man. We got some great shots. Stylistically the film looks like it was made yesterday. It certainly doesn't look old fashioned.

'We worked really hard to get the right shots. At the beginning, it was filmed at night and we had to work in some very dim lighting at Astran's depot. For the interior shots, we had to get the cameraman and the sound man inside, plus their heavy and awkward equipment. It was a very tight squeeze especially if we were filming in the cab shared by Williams and Hook.

'Alec Curtis, the cameraman, was quite a small man so it wasn't so bad for him, but Derek Medus, who looked after the sound, was of much bigger build and the poor bloke had to lie on the top bunk amongst all the drivers' gear.'

The crew had no way of playing any of the filming back and had to rely solely on the talents of Curtis and Medus to produce it properly first time round:

'There couldn't have been a better team. We worked on real time as things were happening. We couldn't plan any shots ahead and we also knew that we couldn't keep sending the trucks back for re-takes due to our film budget and of course the time scales.

'It wasn't until we had finished the filming in Doha and flown home that we could actually sit down in the studio and watch what we had filmed. I remember it was very tense

The Destination Doha film crew. Left to right: Alec Curtis (camera), unknown (general assistant), Tony Salmon (producer), Derek Medus (sound) and Simon Normanton (assistant producer). (SN)

Fuel stop in Austria. Poulton chats to Williams, while Hook shouts to Rivers at the back. (SN)

and nail-biting as we prayed that all the scenes would come out, but of course they did which was extremely satisfying.'

As producer, it was Salmon's job to take the lead and come up with ways to make the documentary as entertaining as possible:

'We had to think everything through and have a plan before we did the shots. Sometimes it worked, sometimes it didn't. Nothing was staged, apart from one bit in the desert, and that was what made it so exiting. We were doing everything for real.

'As we only had one camera, poor old Alec was running around like a madman. At Folkestone, he filmed two trucks driving down the link span onto the boat and then ran back to the other truck and sat in the cab, filming through the windscreen and totally out of breath as it drove on board.

'In Bulgaria we were pulled over by the police and Alec had to lie low and be very quiet in the cab for fear of being arrested and accused of spying with his camera. It was all a bit tense. John Williams played a blinder and distracted the police by jumping out of his cab and talking to them, instead of allowing them to come to him. Meanwhile, true to form, Alec still managed to get a couple of good shots through the windscreen as he pointed the camera without being seen.'

Using the Range Rovers to get the best possible tracking shots of the trucks proved quite difficult at times but with Salmon's previous motoring experience, nothing was impossible:

'I can't believe we got away with some of the action shots. The fact that Alec was sometimes strapped to the back of a Range Rover with nothing more than a belt holding him on caused a bit of stir when we did it, but in reality it was thirty years ago and we got away with murder then. Health and Safety would have a field day nowadays.

'We had to use our imagination and it was a unique way to make films so we just got on with it and managed to produce some amazing results. I can't tell you how good Alec was at his job.

'I never knew what was coming around the next corner and I never knew what was going to happen next. The most exciting thing about it all was that we never wanted to put the camera away because we were in fear of missing something important. In fact there was tremendous pressure on all of the crew to keep filming which made the whole documentary so much more interesting. We had to be with

Alec Curtis (far right) was literally strapped to the Range Rover so he could get the best tracking shots of the convoy on the snowy Austrian roads. (SN)

them every step of the way and get the shots right, too, and of course the rest is history.'

Not only did Salmon have to get the filming right, he also had to capture the personalities of each driver: 'They were all marvellous and sparked off each other very well which made the whole thing very easy for us. Out of the four men, I remember John Williams being a bit of a comedian. You couldn't have written some of his one-liners,' Salmon chuckled.

Normanton recalls that the drivers were all good men and more than helpful whenever they were asked to do something for the film crew. However, there was occasionally friction which caused problems within the camp:

'I can understand that the drivers didn't get on all the time. These were men who were used to working alone and driving at their own pace, so to be instructed to drive in convoy, close together for some five thousand miles must have tested their patience to the limit. They were truly a great bunch of lads and I can't thank them enough for putting up with us.'

At one point in Turkey, Poulton decided that enough was enough and drove off out of sheer frustration:

'I wasn't happy with the amount of time it was all taking so I said to the others that I was going and off I went. I drove about ten miles and then realised that if I carried on I would mess it up for everybody so I stopped and went to sleep. Then John Williams came looking for me. We had a good talk about things and I decided to carry on with the film.'

Tony Salmon had to deal with the problems and keep everyone reasonably happy:

The film crew had to be with the trucks every step of the way as there was no chance to go back for re-takes. (SN)

'I knew I would cause disagreement when I requested certain things, but I was so passionate about what I was doing, I just wanted to make the best film possible and I needed to be brutal at times. In saying that, I too have total respect for the drivers, especially when they drove off the tarmac and out into the desert and then just kept going for miles and miles.

'How on earth they knew where they were heading I'll never know, but of course they were the experts and knew exactly what they were doing. As a film crew, we just tagged along and happened to get some really amazing tracking shots along the way.'

Dave Poulton became increasingly frustrated at the length of time the filming was taking:

'It took about five weeks just to get to Doha. Normally I'd do it in about fifteen days, so you can see my point. Travelling together all the time made things far worse because if one of us stopped we all had to stop whether it was for the film crew, for one of us or for a breakdown. We just seemed to keep stopping all the time.'

At the beginning of the film, Poulton's truck, NVW 484P, is seen to have braking problems which led to he and Williams spending a good few hours stripping down the trailer hubs in an attempt to rectify the problem. To make matters worse, the convoy had only just disembarked off the ferry in Belgium on first day of the journey. At the time the reason for the brake problems hadn't been properly identified so neither Williams nor Poulton knew what they were up against.

The Leyland Marathon also suffered its fair share of breakdowns so the Range Rover full of spares played its part well. Both John Williams and Dick Rivers had their work cut out to keep up with the failing vehicle which never performed well throughout the whole trip. Indeed, it needed major attention on just Day Three of the journey when it broke down in Germany with alternator problems.

However, there were some lighter moments, as Normanton recalled:

'In Austria I remember seeing a microwave oven for the first time in my life when we stopped at a service station. One of the drivers didn't take the cellophane wrapper off his food before he heated it which caused a small explosion in the oven and led to a great deal of excitement in the restaurant. We tried to do it a

Inevitably there were problems and friction along the way. Salmon (right) knew he caused disagreement, but he just wanted to make the best film possible. (SN)

couple of times again for the camera in case it wasn't picked up first time round, but the manager came over and asked us to leave as he didn't want his new oven wrecked!

'I particularly remember stopping at the ski resort of Schladming in Austria. We all got there just in time for the compulsory weekend truck driving ban. It was very much touch and go as to whether we would make it to a particularly good stopping place, but we did ... just.

'The weekend turned out to be excellent. Four English truck drivers in an exclusive resort at the top of a mountain, wearing wellington boots, Parka coats and a variety of hats. It was brilliant for us as a film crew. We got some very funny looks from

Salmon trusted the drivers as they were the experts. The film crew tagged along and happened to get some amazing tracking shots along the way. (SN)

The Leyland Marathon had numerous breakdowns. Here John Williams (left) and Dick Rivers attempt an alternator repair. (SN)

the locals but everyone had great fun and it showed the British sense of humour up perfectly on the film. We spotted a guy skiing on one leg and were all amazed at how good he was - so of course the drivers had to have a go.'

For Salmon the weekend was the highlight of the whole journey: 'The drivers looked so well dressed for a day's skiing. We had so much fun filming that part of the documentary.'

Skiing was not something that Frank Hook would normally have done. He told me he would rather sit in the bar but the film crew wanted to take the drivers up the mountain for more entertainment value so he was roped in with everyone else. 'I have to admit, it was a great laugh,' he said. 'We were all as bad as one another and we kept falling over. We stood out for miles as we looked so stupid.'

Not only did the Leyland Marathon give problems during the trip, but its bad luck must have rubbed off onto its driver as Dick Rivers had an incident of his own at the resort. When he came out of the bar late on Saturday night he fell and smashed his false teeth. According to Poulton, he tripped up on some steps and hit his head. 'Luckily he was drunk at the time and never felt a thing,' Poulton said. The next morning, it was Williams who had the daunting task of fixing the dentures back together again using a new product called Super Glue.

But it wasn't just the trucks and drivers that had problems, as Salmon said:

'I think the worse part of the whole journey for us was at some of the borders, especially when we got past Germany. Not only did the drivers have to do the paperwork and documents for their vehicles and loads, but we as a film crew had a huge amount of our own documentation and it all took a very long time to get processed.

'Normally we would have some sort of hassle wherever we went because officials don't like cameras. It was bad enough taking all our equipment through a big airport in a major city. In this case we were on a commercial route with the trucks, so the border crossings would be much more remote. The officials had probably never seen a film crew before, so they would scrutinise all the equipment which would take an eternity. Sometimes the poor drivers had to wait hours for us to catch up.'

Normanton talked about some of the filming incidents, including how they communicated with the trucks without phones or radios:

The drivers were not suitably dressed for skiing. Nevertheless, they did enjoy themselves. (SN)

'I remember entering Turkey from Bulgaria. Even though I had arranged all the necessary filming and customs permits with the Turkish authorities, the officials on duty at the border post didn't like the film crew and it was very difficult explaining to them what we were doing. The Turks were obviously ready for us when we arrived and we were immediately escorted like a presidential cavalcade with lights flashing all the way from the border at Kapikule to Istanbul.

'We had no idea at all where the officials were taking us and we eventually pulled up outside the Istanbul Hilton Hotel. We were all fed up and tired so decided to stay there the night while we waited for the trucks. At least we got fed and watered properly in the bar.

'On the roads we would arrange to drive ahead by ten minutes then quickly find a suitable parking place, get the camera gear set up and then wait for the trucks to roll by. It was comical really because we'd just about be ready to film and then with an almighty roar the trucks would go past. I swear they never waited the full ten minutes.

'It was so frustrating. We then had to pack the kit away as quickly as possible and spend the next twenty minutes trying to catch them up. It was like the Keystone Cops.'

At night, the crew would book into a local hotel reasonably early, while the drivers stayed in their cabs:

'The truckers used to stay up long after we'd gone to our hotel. I'm sure they were partying, but we had a lot of work to do such as cleaning the equipment and re-loading the film spools for the next day. We put in some very long hours in the hotels and that's why we always looked and felt so knackered the next morning.

'We did socialise on a couple of occasions when we treated the drivers to a slap-up restaurant meal or we all went to a club.

'It was very different when we stopped for the night in the desert as we had no hotel to go to. We stayed up talking and drinking with the drivers then slept out in the open under the stars while they had their luxury bunks. Then it was they who were laughing at us.'

As the convoy rolled on towards Saudi Arabia, everything was going well and luckily none of the trucks had become stricken in the shifting desert sands. Salmon had noticed other trucks which had become bogged down and he thought it would be a good idea to show the viewers how the drivers coped in such a situation.

So, as they neared the tarmac road that followed the Trans Arabian Pipeline south towards Qatar, Tony Salmon asked the drivers to set one of the trucks up to make it look as if it had become bogged down. Once everyone got to the TAP Line, there would be no more chances to get the trucks stuck for the benefit of the film so it was now or never.

The drivers duly obliged and positioned Hook's Scania in a fairly soft section of desert. The Leyland Marathon was then hitched up to Hook's front bumper and, with Dick Rivers tickling the throttle, his truck proceeded to pull Hook cleanly and swiftly out. Alec Curtis was close by to get the perfect shot showing lots of wheel-spin and sand being thrown around. Once Salmon gave the okay the journey continued with the men not even breaking sweat.

Ironically the very next day two of the trucks genuinely did get stuck! At that point, the convoy came across Dick Snow who until then was six hours ahead of them. Snow was also on his way to Doha but was not part of the film. In fact the convoy had first bumped into Snow on his way back to the UK when they drove out through Belgium.

When Snow got back to base, he had taken ten days compulsory time off and then started out again towards Doha. He promptly caught the convoy up – which illustrates just how long the filming was taking – overtook them and then got himself bogged down.

Not wanting to leave their colleague in the intense desert heat everyone helped to dig Snow out and the whole incident can clearly be seen on the film. Unfortunately Snow's truck was stuck up to its axles in the soft sand and no amount of shovelling would free it, so Poulton used a steel bar attached to the front of his truck to try and drag Snow out. All that achieved was to bog Poulton down too, so reluctantly he unhitched and made a hasty retreat while he could.

Normanton said that the incident was taking quite a while to resolve so Curtis was able to move around the stricken trucks to his heart's delight, looking for the very best shots. As he zoomed in on Dick Snow who was crouched under his truck digging furiously with sweat pouring down his face, the following comment was made directly into the camera and microphone by Snowy himself: 'Why the fuck couldn't this have happened outside the Park Royal brewery?'

That was the only line that Snowy spoke and although it caused a huge amount of laughter, it was not kept in the finished film – which many people thought was a great pity.

Before arriving at Doha the convoy broke up. Hook turned off and headed for Kuwait while the others continued south but then stopped along the way so that Williams could repair Astran truck JLL 686K which had been abandoned in the railway yard in Hofuf.

By all accounts a school teacher working casually for Astran at the Doha depot had been driving the truck and abused it until it was useless and had to be towed to the yard and left for repair. Williams was told that the clutch was worn out. However, that seemed to be okay so he and Poulton spent a couple of days under the cab working to get the engine started.

While Dick Snow and the others look on, Frank Hook digs furiously to free Snow's truck. (SN)

Eventually after much verbal encouragement to the truck from Williams, the V8 engine fired up, much to his delight. The film crew caught all the action as Williams brought his beloved truck back to life, speaking as gently to it as if it were a woman.

The narration pointed this up humorously, Benny Green saying, 'For six months, derelict and given the kiss of life in a couple of hours. Between the driver and his lorry, it's a love affair.' The camera zooms in to a close-up of the trailer tow hitch slamming into the coupling on the truck. Normanton wondered if it would be too risqué for the BBC to show on television. Obviously not!

Poulton was the next one to leave the convoy as he headed away to Riyadh. Williams and Rivers then arrived safely in Doha and were filmed talking to John Griffiths, the Astran representative, about their return loads to England. The short but informative sequence was complemented nicely by Benny Green's final comment: 'Destination … Folkestone.'

The film crew celebrated the ending of the filming with a very good night out with the remaining Astran men before everyone went their separate ways. Meanwhile, all the film equipment and the Range Rovers were loaded into Williams's and Rivers's trailers for the trip home.

Simon Normanton summed up the final stage of the journey:

John Williams clings to the mirror of a local truck while instructing the driver to pull. (SN)

'When we all got to Doha, sunburnt, dusty and dirty, we were shown to a hotel by people from the Qatar Ministry of Information who were looking after us there. To the end of my days I shall always remember the sight of the bathroom in my room which had pure white marble and luxurious solid gold taps. It was absolute heaven after five weeks on the dusty road.'

After the filming finished in Doha, the crew flew back to the UK while Normanton went on to Abu Dhabi and Yemen to look into a story about the Queen of Sheba and the Marib Dam. 'Filming *Destination Doha* was such great fun,' he said. 'My appetite for heat, dust and wild adventures are still unquenched, all thanks to that film.'

Back at Astran's Addington headquarters import clerk Bob Haffenden dealt with the documentation of everything coming back to the BBC. 'My claim to fame,' he said, 'is that I was the one who drove the cars back to Henley's hire centre.'

Another Astran driver, Terry Tott, remembers bumping into the convoy on more than one occasion at customs and border crossings. Like Snow, he was also en route to Doha but was not part of the film:

'The film stars were very privileged because they had the BBC crew with them. They were processed at some of the borders far more quickly than the rest of us mortals, so off they went waving at us like royalty with big smug grins. But then we'd catch them up again as they were taking ages to do a particular bit of filming. For some reason they weren't grinning much by then. It was like cat and mouse.'

Like Normanton, Tony Salmon is proud of his film:

'I can be very objective about my own work and I have to say that I rate *Destination Doha* very highly on the list of films which I've made over the years.

'I'm astounded with the standard of photography considering it was made over thirty years ago. I didn't realise the film was as good as this and it really is comparable with anything you would see on the television today. It was ground-breaking at the time and was a genuine road movie. There is something totally different to watch in every scene which keeps the viewer's attention.'

The programme also did Astran a power of good during a couple of lean years. As Bob Paul said:

'We got our own copy of the film and cut it down to a shorter version which we gave to customers in the UK and in the Gulf. It was very successful for us and brought us a lot of new business.

'I was invited by the BBC to the television preview in a special little theatre in London. They treated me like a star. Although *Destination Doha* took an awful long time to film, I have to say it was well worth it and portrayed Astran exceptionally well. I was very pleased with it.'

It was Tony Isaacs who contacted the film's narrator, Benny Green, who first worked for the BBC in 1955. Besides being a jazz saxophonist he wrote and narrated many radio documentaries, particularly about the stage, film, jazz and sport. Green scripted the narration to *Destination Doha* himself – though he hated travelling anywhere!

The drivers were each paid £100 for their parts in the film. Life then returned to normal although they and their Astran workmates did get some ribbing from other Middle East drivers who were looking for autographs and signed photographs.

CHAPTER 16

The Astran Fan Club

Truck enthusiasts are just like any other group of people with a common interest in a particular hobby. Some prefer to take photographs of passing trucks, some visit transport company depots and spend the odd Saturday catching the fleet as it comes in, while others position themselves in lay-bys, roundabouts or on bridges waiting for something special to come past. There are enthusiasts who visit shows such as Truckfest, there are those who collect brochures, magazines and all manner of publicity material.

Some who have more money to throw at their hobby will buy and lovingly restore a full-size truck and trailer. At the opposite end of the scale there are those who build models of favourite rigs for private or public display or to sell on.

Over the years Astran has attracted many fans. When I was a young teenager I wrote to Astran's transport manager asking for details of the fleet, destinations and journeys. Mr Peter Cannon kindly wrote a lengthy letter back and included a couple of Astran giveaways which I still treasure today. When I was researching this book thirty years after writing that letter I finally got to shake Mr Cannon's hand and thank him for starting me off as a true Astran fan!

In recent years enthusiasts' interests in Astran have grown enormously. On internet websites such as eBay photographs of the famous red-and-yellow trucks have been known to command high prices at auction. First editions of Franklyn Wood's book *Cola Cowboys* have recently been sold for over two hundred pounds! When Old Pond Publishing released the DVD of *Destination Doha* in 2009 they found that demand far outstripped their expectations.

Trucks Restoration, Models and Paintings

The following are examples of the tributes that talented enthusiasts have paid to Astran through their truck restoration, modelling and artwork.

David Cassell

David Cassell from Burton upon Trent watched *Destination Doha* on television when he was a student and has had a fascination for Astran ever since. He finally got round to building the 1:24 scale model of one of their rigs in 2008.

The Scania 111 tractor unit is a Heller model kit which had previously been built by someone else. Cassell very carefully stripped it down, a difficult task in itself, and then rebuilt it to Astran spec. Cassell added the roof rack, air-con unit, jerry cans, fuel saddle tank, spade and bucket.

The tandem axle tilt trailer is a modified Italeri model kit with

Keen truck modeller David Cassell has been fascinated by Astran since the 1970s.
(DC)

added fuel belly tank, tackle boxes, water barrel and towing bar. The model carries the authentic Astran registration number UAR 747S.

Russell Hardy

Twenty-six year old Russell Hardy from Sheffield has been building models since he was ten and has his granddad to thank for getting him into truck modelling.

His Mercedes is big at 1:14 scale. It is built from a Tamiya chassis and running gear and has a modified Tamiya Mercedes Benz cab to look like the old-style cab used by Astran. The trailer is based on Tamiya parts and uses wheels, axles and chassis. The bodywork is scratch built.

The whole rig is fully radio controlled with authentic sound for the brakes and horn. Lights are operated via a miniature computer mounted on board the rig. A three-speed gearbox and working diff give the rig its power. The whole lot is powered by 7.2 volt batteries.

The rig measures over 1.1 metres long. Hardy used aerosol spray paints for the main colour and then turned his hand to an air brush for the subtle weathering. The lettering is cut from vinyl and was made in a graphics shop to Hardy's spec. The working model took Russell about a year to get to the stage it is at now.

Jan Dekker

Dutchman Jan Dekker prefers to produce model trucks in the tiny 1:87 scale. Here he used two Scania 1-series cabs from Albedo to convert his model. The second cab was used for stretching the first one.

The tractor chassis is a standard Herpa tag axle rigid chassis, which Jan shortened to the correct length according to Scania specifications. Two huge diesel tanks and the roof rack were scratch built from Evergreen strip and tube. The air-con unit was from Preiser and the sun visor from a Herpa 2-series Scania.

'Plastic surgery' on the cab was done in less than two days. Painting, including some filling and sanding, took about two to three hours and the roof rack another couple of hours to construct. The cab interior is not highly detailed because it is impossible to see much at that small scale.

The trailer was actually harder to build than the Scania. The tilt body is from a new Herpa 13.6 metre trailer model kit, but was placed on a stretched 12 metre chassis. The reason for this is that the newer Herpa chassis are designed for super-single tyres and it is much easier to convert the older 12 metre chassis to a double-wheel configuration. Using more Evergreen tube and strips Dekker scratch built the extra diesel belly tank, tool lockers, rear under-run bar, spare wheel rack, with two spares, and many other little details.

Dekker made the decals himself on his computer and then had them printed by Dutch modeller Piet Oversloot.

Overall Dekker reckons he spent between forty and fifty hours building the model. Searching for reference material, looking for parts in his huge stock of spares, converting and scratch building, designing the decals, priming and airbrushing and the finishing touches like cab top lights and snow chains – all make this Astran replica a priceless model.

Russell Hardy has spent many hours building this impressive radio-controlled model based on an Astran Mercedes Benz. (DC)

Dutchman Jan Dekker specialises in building replica trucks in mini size 1:87 scale. This example is based on Mark Stewart's impressive Scania 141 featured in Chapter 4. (JDE)

Carl Jarman

Carl Jarman from Manchester is an owner driver by day working for a forklift distributor, but at weekends he much prefers to play with his life-size toy. Jarman's pride and joy is the Scania 141 6 x 4 tractor unit built in 1980 which he found in the UK. He instantly fell in love with it. 'It's an ongoing project which my nineteen year old son Paul and I really enjoy working on,' he said.

It is a Dutch spec vehicle with the manufacturer's super deluxe cabin fitted. That includes a quilted engine cover and kerbside mirror fitted to the passenger door, a unique option only found on continental trucks.

Jarman took around six months to get the Scania roadworthy but both he and Paul have spent many more gruelling hours working on it to get it to the condition it is in now:

'The nice thing is that as and when I get some money to throw at the truck we can do extra bits and pieces. That is especially satisfying when we take the truck to shows because people can see we are still working hard to improve the appearance.'

The 12 metre tilt trailer manufactured by German company IWT in 1976 was also sourced in the UK and

The real thing. Carl Jarman's pride and joy is his fully restored Dutch spec Scania 141. (CJ)

came from an exhibition company which had only used it for light duties, so it was in very good condition. Jarman removed the tail-lift, re-worked the rear end and fitted a completely new canvas cover.

Both the tractor unit and trailer are fully liveried in the famous red-and-yellow Astran colours which Jarman said always attracts a huge amount of interest wherever he takes the rig: 'Ninety-nine per cent of people who see it ask me if it is the rig that was featured in *Destination Doha*. It's amazing that they think that.'

Before painting the rig Jarman contacted Hugh Thompson at Astran and asked his permission: 'Hugh was most impressed with what I was doing and had no hesitation in giving me the thumbs up to use the Astran livery. He was very encouraging,' Jarman concluded.

Jarman said that the next jobs for him and Paul were to fit a heavy-duty rear bumper bar, food and tool lockers on either side of the trailer and, if he can afford it, a belly tank for the trailer chassis and an authentic air-con unit for the Scania roof.

Jez Coulson

Truck artist Jez Coulson from Leeds has now become an accomplished amateur painter of commercial vehicles:

'I love painting them all but my favourites are the rigs from the 1970s and '80s, especially the ones engaged on the Middle East run. In particular those from the Astran fleet.'

The fleet's in. Miniature versions line up with Carl Jarman's full-size restored Scania 141 at the Haydock truck show in 2009. (DC)

Jez Coulson has spent around thirty hours painting this intricate collage of Asian Transport/Astran trucks through the ages. (JCO)

317

Coulson used to paint for fun at school and after many years away from the hobby he has started again using water colours but then turned to acrylics as he prefers the quality. He has now been painting more seriously for about eight years.

His latest project, the stunning collage of Astran trucks through the ages, took thirty hours of painstaking work.

A range of Jerry's work can be found on his website: www.jerrycoulson.co.uk

Truck driver Trevor Stringer enjoys painting using a mix of pen and inks and acrylic colours. (TSC)

Trevor Stringer

Truck driver Trevor Stringer from Flitwick near Bedford is also a dab hand when it comes to painting. Stringer aims for the finely detailed, pin sharp finish that can be seen in the painting of two early Astran rigs.

Stringer used a mix of fine-tipped ink pens to achieve the sharp outlines which he filled in with mainly acrylic colours and some water colours for added effect. The painting took Stringer about six months to complete in his spare time and while sitting in his truck during rest periods.

Tony Christie

Tony Christie has only been building model trucks for a year but his results are pretty impressive. Christie runs his own haulage business and drives one of the trucks himself but after major heart surgery in 2009 he was not allowed to drive for a considerable time.

Christie decided to take up model building after one of his drivers built a truck but painted it in the wrong colours. Christie thought he could do better.

Chris Hooper's immaculate 1:24 scale Scania 143 was Christie's first Astran model and as he said:

'I sent Chris some photographs of the finished thing and then he phoned to thank me. He was amazed at what I'd done and couldn't believe that a model could carry so much authentic detail. I was absolutely taken aback with his comments.'

Chris Hooper was suitably impressed with this replica model of his beloved Scania built by Tony Christie. (AC)

Apendix

The Fleet List

10 RMY	Guy Invincible 4 x 2 tractor unit
DRK 798D	AEC Mammoth Major MK5 6 x 4 drawbar outfit
SVB 300F	Scania Vabis 6 x 2 drawbar outfit on loan from Scania dealer
BLH 764B	Leyland Comet 4 x 2 tractor unit on loan from Scania dealer
UHM 25F	Scania Vabis 6 x 2 drawbar outfit
UHM 26F	Scania Vabis 6 x 2 drawbar outfit
YYW 330G	Scania 110 6 x 2 drawbar outfit was re-registered to AMY 147H
WLO 95G	Scania 110 6 x 2 drawbar outfit
BGH 172H	Scania 110 6 x 2 drawbar outfit
ELK 384J	Scania 140 V8 6 x 4 drawbar outfit
ELK 385J	Scania 110 6 x 4 tractor unit
DLP 154J	Volvo F88 6 x 2 tractor unit hired from AVIS
JAN 774K	Scania 140 V8 6 x 4 drawbar outfit
JLL 686K	Scania 140 V8 6 x 4 drawbar outfit
JLB 259K	Volvo F88 6 x 2 tractor unit hired from AVIS
JLB 279K	Volvo F88 6 x 2 tractor unit hired from AVIS
JLU 315K	Volvo F88 6 x 2 tractor unit hired from AVIS
TDX 400K	Scania 110 6 x 4 tractor unit
HMK 501K	Volvo F86 4 x 2 tractor unit
VVU 34?L	Volvo F88 6 x 2 tractor unit hired from AVIS
MUR 501H	Scammell Crusader 6 x 4 tractor unit on loan from Scammell Motors
WTJ 120L	Leyland Marathon 6 x 4 tractor unit on loan from British Leyland
PHK 172M	Scania 140 V8 6 x 4 tractor unit
NTW 562M	Scania 110 4 x 2 tractor unit
WLE 344M	Scania 140 V8 4 x 2 tractor unit
VGF 897M	Scania 110 4 x 2 tractor unit
GUL 588N	Scania 110 4 x 2 tractor unit
GUL 592N	Scania 110 4 x 2 tractor unit
GUL 599N	Scania 110 4 x 2 tractor unit
SYK 280N	Scania 110 4 x 2 tractor unit
KJN 671P	Scania 140 V8 4 x 2 drawbar outfit was converted to tractor unit
KJN 680P	Scania 140 V8 4 x 2 drawbar outfit was converted to tractor unit
KVX 859P	Scania 140 V8 4 x 2 drawbar outfit was converted to tractor unit
NVW 480P	Scania 111 4 x 2 tractor unit
NVW 484P	Scania 140 V8 6 x 4 tractor unit
NVW 488P	Scania 111 4 x 2 tractor unit
NCW 732P	Leyland Marathon 4 x 2 tractor unit loaned/publicity? The only evidence of this vehicle is found in a British Leyland brochure.
PVW 802R	Scania 111 4 x 2 tractor unit
SOO 815R	Scania 111 4 x 2 tractor unit
SOO 816R	Scania 111 4 x 2 tractor unit
SLO 707R	Leyland Marathon 4 x 2 tractor unit on loan from British Leyland
WHJ 193S	Scania 111 4 x 2 tractor unit
WHJ 194S	Scania 111 4 x 2 tractor unit
UAR 743S	Scania 111 4 x 2 tractor unit
UAR 747S	Scania 111 4 x 2 tractor unit)
NKM 284W	Mercedes Benz 1632 4 x 2 tractor unit
NKM 300W	Mercedes Benz 1632 4 x 2 tractor unit
MKN 681W	Mercedes Benz 1632 4 x 2 tractor unit
YTW 875X	Scania 81 4 x 2 tractor unit
AOO 66X	Scania 142 V8 4 x 2 tractor unit
AOO 67X	Scania 142 V8 4 x 2 tractor unit
AOO 68X	Scania 142 V8 4 x 2 tractor unit
GKJ 82Y	Mercedes Benz 1633 V8 4 x 2 tractor unit
GKK 906Y	Mercedes Benz 1633 V8 4 x 2 tractor unit
EKO 948Y	Scania 142 V8 4 x 2 tractor unit
EKO 949Y	Scania 142 V8 4 x 2 tractor unit
EKO 950Y	Scania 142 V8 4 x 2 tractor unit

Other Trucking Titles from Old Pond Publishing

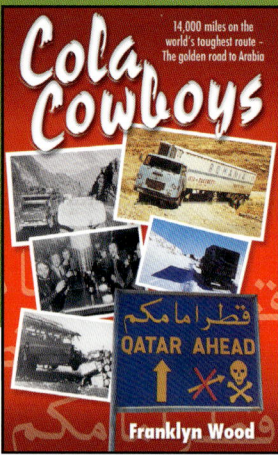

Destination Doha
'The World about Us'

In this 1977 BBC programme meet the extraordinary men who form the 5,000 mile transport link between Britain and the Arabian Gulf. Parts One and Two are both included in this two-disc set. *DVD.*

Cola Cowboys *Franklyn Wood*

The spirited account of what a UK journalist found when he accompanied truckers overland to the Middle East in the early 1980s. *Paperback book.*

Scania at Work *Patrick W Dyer*

This highly illustrated book shows how Scania developed from the L75. This is a book for an enthusiast, by an enthusiast. *Hardback book.*

In the Driving Seat
Alex Heymer

Alex Heymer's driving experience included the RAF (national service), London Transport trolleybuses and buses, Galleon coaches and Gulf Oil tankers. Forty-four years' driving rich in anecdotes. *Paperback book.*

Know Your Trucks
Patrick W Dyer

'This is an excellent little book and does exactly what it sets out to do: that is, to give enthusiasts the basics on which to begin to build their interest in trucks and the trucking industry.' *Trucking* magazine. *Paperback book.*

Truckers North Truckers South
Leslie Purdon

In 1945, Shay leaves school to become a trailer-boy on an Atkinson truck and is taught to love life on the road by the driver, who knows all the tricks of the trade. *Paperback book.*

Juggernaut Drivers
Leslie Purdon

A light-hearted read for everyone who likes trucks and trucking, featuring a fictional gang of 1970s owner-operators. In good times they run legal; in bad times they cut corners. *Paperback book.*

Free complete catalogue:

Old Pond Publishing Ltd, Dencora Business Centre, 36 White House Road, Ipswich IP1 5LT, United Kingdom

Secure online ordering:
www.oldpond.com

Phone: 01473 238200 Fax: 01473 238201